AF407484

ADAPTERS

Turning Experience into Advantage in the Age of AI

A practical framework for mid-career professionals navigating the AI transformation

Book By

FELIPE VEIT

NEVENTRY LABS
PUBLICATIONS

ADAPTERS: *Turning Experience Into Advantage In The Age Of AI*
First edition | Cover design: *Mubi Designs* | Edition, Formatting and Interior Design: *Rakib Hossen* | Proofreading: Samantha Khan | Printed in the USA.

Library of Congress Control Number (LCCN): 2026910014
ISBN 979-8-234-06186-7 (hardcover)
ISBN 979-8-234-06150-8 (paperback)
ISBN 979-8-257-30401-9 (Kindle ebook)
ISBN 979-8-234-06215-4 (ebook)
ISBN 979-8-234-06222-2 (audiobook)

Library of Congress Subject Headings
Career development, Artificial intelligence, Technological change, Occupational change, Professional adaptation, Future of work, Personal transformation, Self-management.

Praise for ADAPTERS

"The playbook professionals need for the AI era. Veit outlines with clarity how to cultivate the human edge no algorithm can replicate, turning years of accumulated experience into the very advantage that sets you apart."

Nannis GadelRab, President - Mondelēz Canada

"What makes *Adapters* especially relevant is its practical view of adaptation. As companies continue to invest in AI learning, experimentation, and skills development across the workforce, this book offers an important perspective: technology creates the most value when people are equipped to grow with it. Veit makes a strong case that the future belongs not only to those who adopt new tools, but to those who build the confidence and capability to use them well."

Sudeep Banerjee, Sr. Vice-President, Procurement, Beauty & Wellbeing - Unilever

"Adapters is not just about AI, it's about what it means to adapt and lead through change. I've seen Veit's approach in action, both as his friend and running teammate, watching him set ambitious goals that pushed his own limits and inspired everyone around him. 'Adapters' embodies that same discipline, curiosity, and resilience that made him such a strong competitor. It's a practical and compelling guide for leaders ready to reinvent themselves and transform experience into competitive advantage in the age of AI."

Vander Freitas Jr., Sr. Vice-President, Global Supply Planning - The Estee Lauder Companies

"AI is reshaping how work gets done, and staying relevant requires real adaptability. But adaptability takes discipline and courage: making time to learn, testing new tools, and adjusting as technology evolves. It also requires resilience in the face of setbacks and uncomfortable learning curves. Adapters provides the practical framework professionals need to navigate this transformation successfully."

Thiago Marchi, Senior Director Operations, Head of Global Capability Center México - The Kraft Heinz Company

"If you know Felipe, you know he's a genuine, thoughtful human being. That shines through in his writing. What I love about Adapters is its focus on the mid-career professional, who may feel most at risk as AI potentially threatens the landscape. Felipe provides a practical, human way to adapt, stay relevant, and move forward without starting over."

Todd Caponi, Author of The Transparency Sale, The Transparent Sales Leader, and Four Levers Negotiating - President, Sales Melon LLC

"One of the strengths of *Adapters* is its balanced view of AI and people. Veit explains that the future of work is not just about using new technology, but also about helping people build the skills, confidence, and good judgment to use it well. This is a useful and timely message for organizations and professionals dealing with fast change."

Marcelo Scarcelli, Executive Vice-President, Business Transformation - Unilever

"What makes this book compelling is not only Veit's path, which I had the pleasure of working closely with in the past, but also its relevance for executives navigating reinvention in the age of AI. It is a thoughtful and practical reflection on how to adapt, learn, lead, and evolve in a changing world."

Alexandre Martinez, Sr. Vice-President,
LA. Finance - Barry Callebaut

"At a moment when many executives are still working out where to apply AI, boards have a real role to play: sponsoring the work, asking the right questions, and protecting what AI cannot own, like judgment, trust, and the relational work that makes good leadership more than an algorithmic summary. ADAPTERS takes that angle seriously, written with the clarity of someone who has done the work. A useful read for directors thinking about where AI fits for their executive teams, and where it does not."

Alexandre de Toledo Corrêa - Board Director and
General Manager with experience in the CPG, Technology, Nanotech,
and Chemical sectors

"Having known Veit for over fifteen years, since his early days at Unilever in Brazil, I've watched his remarkable story from humble beginnings to a global leader in AI. In Adapters, he transforms that experience into a powerful and timely message: in an age shaped by AI, those who adapt will thrive. This book is both an inspiring story and an actionable roadmap for anyone looking to stay relevant in a rapidly changing world."

Carolina Singer, Global Head of Sustainability and
Responsible Business - Unilever

"What separates thriving companies from those left behind isn't technology or capital, it's adaptability. Felipe has created a comprehensive guide to building this critical capability across every level of an organization. 'Adapters' delivers actionable strategies that work in real-world business environments. This book belongs on every executive's desk."

Nelson Teixeira, Supply Chain Planning Officer -
Mondelēz International

"Adapters is smart, practical, and refreshingly clear. Felipe shows how leaders can navigate AI with confidence, judgment, and human insight. The ADAPT framework turns disruption into opportunity through actionable steps and clear examples. This is a guide leaders can put to use right away."

Joleen Liang, CEO and Co-Founder - Squirrel AI

"Felipe's journey is a powerful example of how professionals can reinvent themselves with courage, curiosity, and purpose. This book offers valuable inspiration for mid-career leaders looking to embrace AI as a catalyst for growth and transformation."

Alexander Pose, President -
AGP Mineral Solutions

"True relevance doesn't come from competing with technology, but from deepening what makes us human. Adapters is not just a book about AI, it's a reflection on identity and the courage to evolve without losing yourself."

Carolina Ludwig - Human Behavior and Emotional Reprogramming
Strategist

"Few people I know have embodied reinvention as genuinely as Felipe, not as a strategy, but as a way of showing up. Adapters is the honest distillation of that lived experience, making a compelling case that your experience is actually your greatest edge."

Jin-Yan Ang, Procurement Director - Unilever

"The value of Adapters is that it treats adaptation as a business discipline, not a buzzword. Felipe shows that in a time of fast technological change, the winners will be the people and organizations that combine learning agility with human judgment."

James Kyle, CEO and Founder –
Millennium Group Broadband Network Infraestructure

"Finally, a book about AI that focuses on execution rather than prediction. Adapters gives professionals the practical steps needed to reinvent their careers and stay relevant today. During our time together in the master's program, Felipe had a real impact on how I approached AI, making it feel more practical, accessible, and relevant."

Eddie Heffernan, Sr. Business Development Director -
SMS Sysco Corporation

To my wife and son, and those who choose reinvention over resignation.

"We shape our tools, and thereafter our tools shape us."

\- Marshall McLuhan

Table of Contents

Preface

The Hive Moves. So Do You.

During the fourteen months it took to research and write this book, I spent a good amount of time reading about how living systems respond to change. There was one documentary called *A Civilization 100-million-Year-Old* about honeybees, and something about it reframed everything I thought I knew about enduring change.

The lesson I took from that film was not simply their resilience across several critical moments in Earth's history, but the way they navigated each one. Honeybees have existed for roughly one hundred million years. Across thousands of species, they have developed ways to organize work, share information, adjust roles, and make collective decisions that have enabled them to persist across radically different environments. A hive is not static. It depends on coordination, specialized roles, constant feedback, and a remarkable capacity to respond when conditions change.

That image of the hive mattered to me for another reason. When a colony can no longer remain where it is, it does not move without direction. Scout bees go first. They explore, assess, return, and communicate what they find through the waggle dance, a movement that encodes direction, distance, and quality as information the rest of the colony can rely on and act on. When conditions turned against them, it responded dynamically. The scouts do not wait for certainty. They go early, report back, and trust the hive to follow when the signal is clear.

That is why the honeybee became the symbol of the ADAPT method, and of this book. In a sense, I have tried to do what the scouts do. I went first, not because I was certain of where I was going, but because I was willing to move before the answers were clear. This book is what I brought back.

That scout mentality shaped how I approached both my career transformation and the writing of this book. The research involved reading across technology, labor economics, organizational behavior, and innovation, as well as something more practical: experimentation in my own career. For over 25 years, I worked in procurement and supply chain roles, much of that time at Unilever in Brazil, Switzerland, and the USA. Writing was never part of the plan. What I had was a story I needed to tell and a set of tools that made telling it possible.

Writing is something I have never fully let go of, not as a steady practice, but as an instinct that never disappeared. Career came first, followed by all the other layers of life, and writing became one of those things I kept postponing for many years. AI changed that, not by replacing the work, but by making it possible alongside a demanding job, a master's degree program, and a family that deserved more of my time than I was often able to give.

I should be direct about how AI shaped this book, since it's about working with AI, and transparency on that point matters to me. I used several AI systems throughout the project for brainstorming structure, stress-testing arguments, broadening research, and refining prose when a paragraph wasn't working, but I couldn't yet see why. The same tools helped me build the companion platform, adapterslab.ai, and prototype one of the products described in Chapter 7.

One of the strongest lessons from this process was that AI without human oversight produces poor results. I tested this directly. When I let AI generate entire sections without heavy guidance, the output was generic, shallow, and occasionally wrong. It sounded confident while saying nothing. AI sped up the process and gave me capabilities that would normally require a team, but it didn't make the work easier. The hardest parts were entirely human problems that no tool could solve: finding the emotional truth of a story, deciding what to cut, staying honest when the writing wanted to become something more flattering, and keeping a single argument coherent across sixteen chapters.

But there's a more personal reason this book exists, and sharing it requires vulnerability. At one point in my career, approaching my fifties, the exhaustion was real. After an entrepreneurial venture that collapsed during COVID, returning to the corporate space felt like the only available path. Everything built around that venture, the savings poured into it, had been lost. The pride that went with it took longer to recover. The work gave something to hold on to. The feeling of having fallen short, and not being ready to admit it, came along regardless. The experience of twenty-plus years no longer felt like an asset. It felt like something already obsolete.

What changed was a combination of personal stubbornness and circumstances that were more generous to me than I had any reason to expect. Unilever gave me something I will never be able to fully repay. Not just a job. They gave me freedom. Freedom to experiment with AI tools before most of my colleagues and even the market saw the need. Freedom to fail in front of people and try again. Freedom to run workshops, build prototypes, pitch ideas through an internal competition, and join the creation of an AI Champions network that connected me with people dealing with the same questions I was. Leaders in the organization saw potential where I sometimes saw doubt, and they backed it with real support by providing resources, time, encouragement, and, most importantly, trust.

Not every organization would have done that. I know that. And I know that my story is shaped by that environment in ways that not every reader can replicate. But the principles underneath are portable. The willingness to explore, the discipline to keep learning, and the courage to act before feeling ready do not belong to any one company.

This book is my way of giving back. Not to Unilever specifically, but to every professional who is on a path like the one I was on a few years ago. I wrote it for the leader who sees AI arriving and does not know where to start. For the teacher wondering what automation means for her classroom. For the supply chain analyst, the marketing coordinator, the finance professional, the nurse, the salesperson, and anyone who has spent years building expertise and now senses that the rules are changing before their eyes. The goal is not simply to motivate. What matters more are tools and frameworks that help people act.

That is why the book introduces the ADAPT method, a practical approach to career reinvention built around five principles: Awareness, Direction, Action, Positioning, and Tech Empowerment. These principles are not theoretical. They emerged from my own research and experiments, as well as from patterns I observed across organizations learning to work with AI.

You will notice that many of my examples come from procurement and supply chain work. That is the world I know best, and I see no reason to pretend that lens does not exist. I have deliberately expanded the illustrations across sectors, including marketing, healthcare, education, technology, and others. The lessons of adaptation are not limited to one profession. But it was in procurement that my thinking was shaped. Instead of stepping away from it, I have tried to translate its lessons into something broader, a way of approaching change that readers from any field can use and relate to. If you are reading this from a completely different background, the ADAPT method does not require a background in procurement. It requires a willingness to see where you are.

Beyond the Printed Page

There is one limitation to books that bothered me throughout this process. A non-fiction book captures a moment in time. Unlike a living system, it cannot update once it is printed. In a field evolving as quickly as artificial intelligence, that limitation becomes a real problem.

To address it, I created adapterslab.ai, the living companion to this book. The platform extends every chapter into an interactive, continuously updated experience. It gives readers a place to explore the ADAPT frameworks in real time, run guided exercises, and generate practical outputs such as adoption plans, role-based workflows, and personalized checklists. When platforms evolve, features update, or new best practices emerge, the reader sees the latest version without waiting for a new edition. While the book provides the narrative, the mental models, and the why, the platform provides the living layer of the how and the what now.

If you are holding this book because you are curious about AI, I hope you leave with more than curiosity. If you are holding it because you are anxious about your career, I hope you leave with a plan. And if you are an experienced professional in any field, carrying the weight of having done a lot and not knowing what comes next, I wrote this for you.

I was you. And I am here to tell you that the best chapters of your professional life might still be ahead if you are willing to adapt.

Introduction

The Rules Changed. Your Experience Still Counts.

You are good at your job. You have spent years, maybe decades, building expertise, earning trust, and delivering results. You know your industry and your craft, yet lately, something has changed.

It shows up in small moments. A junior colleague automates a report in an afternoon that once took your team a week. A meeting fills with terms like "prompt engineering" and "agentic workflows" that you only half understand. You see a demo of an AI tool that handles a meaningful part of your work, only faster. You read another headline about AI changing your industry and feel a flicker of something you cannot quite name.

That feeling is not a crisis of confidence but a matter of relevance. The rules of the professional world are being rewritten, and the skills that got you here may not be the ones that carry you forward. The real fear for experienced professionals is not that a machine will take their job tomorrow. It is the more gradual possibility of becoming less essential in the job they already have.

A 2025 Gallup workplace study found that frequent AI use in professional settings continued to rise sharply through the end of 2025, with knowledge workers reporting the most significant changes to their daily routines. Microsoft's 2025 Work Trend Index described the emergence of what it called the "Frontier Firm," organizations restructuring around AI-augmented workflows, and warned that professionals who delay adaptation risk finding themselves in roles that no longer exist in their current form.

You are not behind. You are not too old. And you are not alone. The fact that you picked up this book means you have already done the hardest thing: you have decided to pay attention.

Who This Book Is For

Most conversations about artificial intelligence focus on the technology: models, parameters, billion-dollar valuations, and the race between companies and nations. Those conversations matter. But they miss the human story underneath.

The deeper story I want to explore is about people. Specifically, it is about professionals who have spent ten, twenty, or thirty years developing skills, building reputations, and planning for futures that assumed the rules would stay roughly the same.

For early-career workers, the AI era tends to feel more like an opportunity than a threat. They are still forming their professional identities and tend to adopt new tools quickly, without much attachment to what came before.

But for professionals in the middle, roughly between their early thirties and mid-fifties, the experience lands differently. This is the stage of life when the stakes are highest. Mortgages are running. Children are in school. Career plans have been built on the assumption that expertise accumulated over the years will continue to grow in value. And then, over the course of a few months, you start hearing that a tool can draft the report you spent a decade learning to write, analyze the data you were hired to interpret, or build the presentation your team spent three days assembling.

For senior executives and professionals closer to the end of their careers, the disruption may feel more distant. Their roles are anchored in strategy, governance, and relationships, capabilities that AI augments but does not easily replace.

The result is an anxiety that millions of professionals are experiencing right now but not saying out loud. This book is written for them. For the manager who sees AI tools arriving and does not know where to start. For the teacher who hears about automation and wonders what it means for their purpose. For the supply chain analyst, the marketing coordinator, the finance professional, the nurse, the auditor, and anyone who has spent years building expertise and now senses that the rules are changing.

Reinvention Does Not Require Starting Over

There is a popular narrative about career reinvention. It tends to center on a twenty-six-year-old dropping out of a graduate program to launch a startup, or a tech prodigy pivoting from one venture to the next.

That narrative is real, but it is not the whole story. The reinvention that matters most, the kind that is hardest, happens in the middle of a career. It happens after setbacks. It happens when the path you walked for two decades forks without warning, and you must decide which direction to take while carrying everything you have already built.

I know this because I lived it. Before writing this book, I spent more than two decades in procurement and supply chain roles, most of that time at Unilever. I moved across countries and industries, survived an entrepreneurial failure, rebuilt from significant lows, and eventually reinvented myself as an AI Product Strategist, a role I could not have imagined two years earlier.

What I discovered through that process and working with professionals across organizations is that the people who manage these transitions successfully are not the ones with the most technical knowledge. They are not the youngest or the most comfortable with technology. They are the ones who learn to treat their accumulated experience as raw material rather than baggage, and who develop a repeatable way to respond to change rather than react to it. That is what this book is about.

The ADAPT Method

Over the past three years, through my own reinvention and through working with professionals facing similar crossroads, I began to see patterns. Not everyone who adapted did so the same way, but the people who adapted well tended to do five things consistently.

The first was **Awareness**, the discipline of noticing changes in their environment before those changes became a crisis. Waiting for a reorganization memo was not part of their approach.

The second was **Direction**, not a perfect destination, but a working sense of where to aim. Certainty was a luxury most could not afford, and the ones who moved forward accepted that early.

The third was **Action** through small, low-risk experiments rather than grand plans. Testing an AI tool on one real task. Volunteering for a pilot. Building a prototype over a weekend.

The fourth was **Positioning**, making growth visible to others. Sharing what was learned, writing about it, and teaching a colleague.

The fifth was **Tech Empowerment**, a working relationship with the tools reshaping their field. Not chasing every new release but building a practical stack that made daily work faster, broader, and more creative.

Those five moves became the ADAPT method. It is the framework at the center of this book, designed for people who need to evolve their careers without abandoning the expertise they spent years building.

Before we reach the method, we start with something equally important: the Polymath Approach. In an era when AI handles depth with extraordinary speed, the people with the greatest advantage are those who can connect ideas across domains, bringing breadth, pattern recognition, and lateral thinking that comes from having worked in more than one world. If your career has been nonlinear, if you have changed functions or industries or worn too many roles to summarize in a single job title, that is not a weakness in the age of AI. It is your edge.

What This Book Is, and What It Is Not

This is not a technical manual. You do not need a background in computer science, data engineering, or machine learning to apply anything in these pages. If you can write a clear email, you already have the foundational skill that matters most when working with AI: the ability to articulate what you want.

This is not a prediction book. I will not tell you which jobs will disappear or which industries will dominate. Nobody can make those predictions with any real certainty. What I can offer is a method for responding to change as it happens, a way of staying in motion when the rules keep rewriting themselves.

This is not a motivational book. Motivation is useful, but it is a temporary feeling. What lasts are habits, systems, and the kind of evidence you generate by doing things rather than reading about them. Every chapter in Part II includes exercises and experiments designed to produce that evidence, and all of them can be completed at the adapterslab.ai platform.

What this book is, at its core, is a practitioner's guide written by someone who lived the transformation before writing about it. I wrote it between conference calls, on airplane tray tables, and in the early morning hours before my family woke up. If the prose occasionally reads like it was written by someone who was learning to write a book while writing one, that is because it was. But the ideas are road-tested because I tested them on the road.

How This Book Is Organized

Adapters is built in three parts, and each serves a different purpose.

Part I: From Shock to Shift is a memoir. It begins with a late night when I first tested an AI tool on a real work task and understood, unmistakably, that the expertise I had spent twenty-five years building was about to mean something different. From there, it reaches back to my childhood in Brazil, the early jobs, the international moves, the years of professional stagnation I call the 'hidden plateau', an entrepreneurial venture that collapsed during the pandemic, and the return to corporate life that ultimately led to reinvention. I tell this story not because it is extraordinary, but because the pattern inside it, stability, disruption, adaptation, is universal.

Part II: The ADAPT Method is your framework. It opens with the Polymath Approach, a mindset chapter that reframes breadth of experience as a strategic asset, then moves through the five steps of ADAPT, each with case studies, exercises, and practices you can begin this week. Along the way, you will meet other Adapters: professionals from healthcare, education, small business, and other fields who navigated their own versions of AI-driven reinvention.

Part III: Working the Future zooms out from the individual to the system. It opens with a narrative walkthrough of an AI-augmented workday, then confronts the real limitations and failures of AI, because a credible book about embracing these tools must also be honest about where they break. From there, it explores how AI is changing roles across industries, examines the human capabilities that remain irreplaceable, offers a practical playbook for teams and organizations working through adoption, and closes with a chapter that returns to the book's core themes and issues a challenge for your first seven days.

This book is also a living document. The world of AI changes fast, and any printed page risks becoming outdated. The adapterslab.ai platform extends this static content with updated tool recommendations, new case studies, and interactive exercises to continue your journey.

The Opportunity Inside the Disruption

Every major technological transition in history has produced a period of anxiety followed by a period of expansion. The printing press, the steam engine, the personal computer, and the internet. Each one unsettled old certainty and created new possibilities that the people living through the disruption could not yet see. We are in that period now. And the professionals who will benefit from what comes next are not the ones who predicted the future perfectly. They were the ones willing to explore it early.

In nature, adaptation is not a one-time event. It is an ongoing process of reading the environment, experimenting with responses, and keeping what works. The scouts do not wait for consensus before moving into new territory. They go first, test, and report back. The hive follows.

This book is written for the scouts, the professionals who are ready to move before the path is fully clear. It is also written for those who are not yet ready but want to be.

The rise of artificial intelligence does not mark the end of human relevance. But it does mark the end of passive careers that coast on accumulated expertise without continuing to evolve. The professionals who thrive in the years ahead will not simply react to change. They will learn to move with it, to shape it, and to find in it the foundation of what comes next.

They will become Adapters. And that journey begins here.

PART I

FROM SHOCK TO SHIFT

Traces the personal path that led to this book. It begins with a moment that made clear the meaning of expertise was changing, then moves through the experiences that shaped what came next. This is not a story about an exceptional life. It is about a pattern most professionals will recognize: the rules change without notice, the old map stops working, and you must keep moving anyway.

1

WHEN EVERYTHING CHANGED

The AI Shock (2022)

It was 2022, late at night in my home office in Orlando, Florida. My family was asleep, and my senior terriers were curled up at my feet, just as they always were when I worked late. I had recently heard about a new AI tool called ChatGPT. It was still in beta, not yet released to the public, and the excitement around it was mostly confined to tech circles I did not usually follow. I had managed to get access and decided to test it on something real.

After a round of initial experiments, I uploaded a dummy contract and asked for revisions to the points that would represent liabilities in the document. The response came back in seconds. The accuracy was impressive, and one observation got my attention: it flagged a renewal clause I probably would have missed, especially that late at night when my focus was already fading.

Curious to push further, I asked it to draft a professional email for a supplier negotiation scenario I had made up. Normally, I would spend half an hour polishing the tone and balancing firmness with diplomacy. This time, in seconds, I had a draft that was not only polite but persuasive, with leverage points I had not thought to include. It was not perfect, but it gave me a strong starting point, one that would have taken me much longer on my own.

For a long moment, I just sat there staring at the screen. The house was silent, but my thoughts raced. Work that had taken me more than twenty years to master could now be done by a machine. Fast. Competent. Not flawless, but close enough to matter.

What I did not fully grasp at that late-night hour was how quickly ChatGPT would break into the mainstream. Within weeks, the tool was the subject of a major feature in The New York Times titled "*The Brilliance and Weirdness of ChatGPT*," which captured the shock and excitement coursing through the tech and business world. The article framed ChatGPT not just as a clever toy but as a genuinely disruptive force, one that could reshape how people wrote, searched, and thought about knowledge work.

I realized that things were changing in a meaningful way, and my career was just one part of my life that would be affected. I did not feel like everything was ending, but I could see this was a turning point. Perhaps the most disruptive I had ever witnessed. I knew I could not ignore it.

This was the moment it became clear that continuing to work the same way was no longer a safe option. At the same time, something else came into view: a new path, one where tools like ChatGPT could fundamentally reshape how work got done. The answers were not all there, and predicting what came next was not the point. But the readiness was. That readiness came from somewhere specific. Growing up in Brazil, you learn early that stability is not guaranteed. You either move with what is coming, or you watch everyone else move ahead. A whole life spent adapting had made that instinct second nature. This was not the moment to stop.

That feeling, the mixture of awe and unease that comes from watching a machine perform a task you spent years learning to do, is not unique to procurement. A radiologist in Chicago described the same sensation when an AI flagged a tumor she had missed on a scan. A senior copywriter in London felt it when an AI produced a brand tagline in ten seconds that her team had spent three days developing. A financial analyst in Singapore recognized it when a model generated a market forecast that matched his own but took minutes rather than a week. The specifics vary. The feeling is the same. And how you respond to it, whether you freeze, retreat, or lean in, is the first real test of adaptability.

Roots of Adaptation (1980's in São Paulo)

I grew up in Guarulhos, a working-class city outside São Paulo. You may recognize that name because it is where Latin America's largest international airport, GRU, is located. Our modest house faced a cemetery. Its terracotta roof was often cracked from kids climbing and playing on top. Dangerous as it was, nobody stopped us. From up there, I could see the funerals and the grieving families who gathered on weekends and holidays to honor their loved ones. The cemetery was a place of sadness, but for children like me, it was also strangely alive. My friends and I climbed its fruit trees every day to pick blackberries, jambos, and peaches. The rows of graves were just part of the playground. The only rule was to get the fruit before the guards, who were armed with German Shepherds and Rottweilers, caught us.

This was our world in the 1980s. While rich kids had Atari video games at home, we had the streets, that cemetery, and the rooftops. The fanciest technology in our house was a 14-inch Mitsubishi TV and a small radio that barely captured any signal. Staying indoors was not really a choice; there was virtually nothing to do inside our small houses anyway. We were pushed outside, into the real world, dealing with whatever came our way. When I was inside, I would spend hours next to that radio, listening to adult programs and sometimes hearing things a seven-year-old probably should not. More than anything, I waited for the Palmeiras commentaries. To this day, Palmeiras is one of Brazil's biggest and most successful soccer clubs, and my passion for them started the day my father first took me to the stadium at four years old.

During holidays, the cemetery got crowded, and parking became impossible. Cars filled every street around our house. That is when the *flanelinhas* appeared. These were men who approached drivers and offered to watch their cars, a questionable promise of protection against robbery. They got their name from the cloth, flanela, they carried, waving it to get attention and wiping windshields to earn their keep. Most drivers would agree and hand them some change, not really for protection but out of habit and generosity. Times were rough in Brazil. Flanelinhas were, and still are, largely part of Brazil's informal economy, ordinary people making a living when real opportunities were hard to find.

At seven years old, watching this happen from the roof every weekend, I decided to try it myself. Our situation at home was tight. My father woke up at 3:15 AM every morning to catch public transportation across São Paulo, arriving at work by 7:00 AM and not coming home until 9:00 PM after six or seven hours of commuting. My mother ran a small shoe shop out of our garage to help pay the bills. I do not recall seeing more than a handful of customers in the time she had it.

When I told my parents I wanted to watch cars to earn money for my own AM radio, one that worked so I could follow Palmeiras properly, they did not say no. They understood our circumstances, and their generation believed that the earlier a man started earning his own money, the better.

My plan was simple: face the adult competition, ask the cemetery visitors to watch their cars, earn a few Cruzeiros, and buy my radio. On my first day, by 6:00 AM, I was already in front of my house claiming my spot. The first car I approached was occupied by an older couple. The man looked straight through me. The woman looked at me with concern and asked whether I had parents, whether I was hungry, and whether I was attending school. I had not expected that and told her the truth: I wanted to make money to buy a radio and help my parents. She hesitated, then agreed. When they came back from the cemetery, she handed me two Cruzeiros and a lollipop. Knowing how hard my parents worked just to keep us going, that sum, probably less than a dollar in today's money, felt like a fortune.

That first success kept me in that spot for the rest of the day. The responses were all over the place. Some people agreed to let me watch their car, but walked away at the end without paying anything. Others, like that man, looked through me entirely, treating me like just another street kid they did not want to deal with. But then there were people like that woman, who showed some humanity. She paid me, but more than that, she took the time to ask questions, to see me as a person rather than a problem. Each interaction taught me something different about people, about how the world worked, and about what I would need to learn to survive in it.

That day left a mark. Life in 1980s Brazil was hard, and survival meant creating opportunities where none existed. Watching cars might sound like nothing, but to me it meant opportunity, pride, and a first real taste of making something happen on my own. It was also my first lesson in adaptation: when the world gives you nothing, you find a way forward.

This is a pattern many professionals recognize later in life: the moment you realize that the world does not always provide a clear path, and that the most valuable opportunities are often the ones you must create for yourself.

The First Work Revolution (Securit, 1994 to 1997)

In Brazil during the 1990s, it was almost a requirement for boys my age to get real jobs in corporate environments. Fourteen was the minimum working age, and society expected you to start contributing right away. Even before my fourteenth birthday, I was already reading the Sunday newspapers, scanning the classified sections where office boy job ads took up entire pages. Not small notices. Whole pages are filled with companies looking for teenagers to join their workforce.

In 1994, as soon as I reached legal working age, with the help of a neighbor, I got a job at Securit, an office furniture company in Guarulhos. Office boy was one of the most common entry-level jobs in Brazil at the time. São Paulo's streets were full of teenagers like me carrying envelopes and documents from building to building. We were the link between companies in a city that still ran on paper and hand-to-hand delivery. It was not just a job. It was an entry into the adult world, and everyone understood that.

FOLHA DE S.PAULO

DOMINGO, 12 DE DEZEMBRO DE 1993

EMPREGOS

OFFICE BOY Precisa-se, 14-17 anos CR$ 150/mês Tel: (011) 257-4832	**Secretary** Typing skills CR$ 400/mês Av. Paulista, 502	**Warehouse Assistant** Physical work CR$ 280/month Tel: (011) 492-5637
Bus Driver License required CR$ 500/month Fax: (011) 883-2156	**Receptionist** Bilingual preferred CR$ 300/mês Tel: (011) 876-3294	**OFFICE BOY** Entregar documentos CR$ 150 R. São João, 324
OFFICE BOY Urgente, início imediato CR$ 150/mês R. Augusta, 1247	**OFFICE BOY** Horário flexível CR$ 180/mês R. Consolação, 89	**Janitor** Night shift CR$ 220/month Tel: 567-8234
Sales Assistant Experience required CR$ 250/month Tel: 284-9103	**OFFICE BOY** Courier/office boy CR$ 170 Tel: 341-8765	**OFFICE BOY** Treinamento fornecido CR$ 160 Fax: (011) 738-4521
OFFICE BOY Meio período CR$ 120/mês Fax: (011) 225-7841	**Cashier** Retail experience CR$ 320/mês Fax: (011) 654-2198	**Data Entry Clerk** Computer skills CR$ 350/mês Av. Ipiranga, 1105

Figure 1 - A Reproduction of How the Jobs Classifieds Looked in Brazil in the 1990's. Made with Nano Banana PRO (2026).

The routine was brutal. I woke at 5:30 AM, caught two crowded, stinky, noisy buses for an hour, and arrived in time for the free company breakfast: a piece of bread with butter and a cup of black coffee, sometimes with a hint of milk. By 7:45 AM, I was already on duty. I worked until 5:30 PM, then went straight to school, where I sat in class until 10:45 at night, fighting exhaustion. All for a monthly wage of what would equal about 120 U.S. dollars in today's money.

The job looked easy, but carried real risks. Vale do Anhangabaú, nearly two hours away by public transportation, sat close to São Paulo's central station, where the nation's major financial institutions had their headquarters. It was one of the most dangerous areas for office boys, and one we could not avoid since that is where most of our deliveries ended up. Banks, investment firms, company headquarters, government offices: all concentrated in the same area. Walking through those streets with envelopes full of contracts, checks, and important documents made you a target. Criminals knew our routes better than we did.

On my very first day, I was robbed under the famous bridge. Four young men surrounded me before I had time to react. My Champion watch was gone, along with the coins and bus passes in my pocket. What they never found was the company's cash, carefully packed in an envelope hidden in my underwear. Shaken but determined, I took the long bus ride back to Guarulhos, gripping that envelope the entire way. Protecting the company's assets mattered more to me than my own safety. That was my idea of work ethic at fourteen.

The danger did not end once you left the financial district. A few weeks later, I was sitting in the last row of an almost empty bus when a guy around my age sat right next to me. The bus had plenty of empty seats, so I knew something was wrong. He pulled out a knife, kept it low so other passengers could not see, and asked for the brown envelope I was carrying.

My reaction was almost automatic. Instead of panicking, I opened the envelope and showed him what was inside: just papers and contracts, nothing of any monetary value to someone on the street. I told him it was worthless, just documents I needed to deliver. Then, without really thinking about it, I reached into my pocket where I had a bus pass and a piece of bubble gum. I offered them to him as if we were friends, as if this were a normal conversation.

The change was instant. He went from threatening me to looking almost confused. He took the gum and the bus pass, stood up, and said "You were lucky" before getting off at the next stop.

What had happened, without my knowing it at the time, was my first negotiation instinct in action. Instead of treating him like an enemy, I had shown him the reality of my situation and offered something small but genuine. It worked because I had not tried to lie or fight. I had just been direct and human.

Inside the office, the danger of the streets gave way to monotony and a different kind of pressure. Olivetti typewriters clattered nonstop. Carbon paper was our copy machine. Since Xerox barely existed in Brazil at the time, every document required a duplicate made with those messy sheets, leaving hands permanently smudged black by the end of the day. Memos traveled hand to hand, invoices were copied line by line, and signatures were collected in person. In this world of manual processes and analog communication, I was not just an office boy. I was part of the living infrastructure of a company untouched by the digital age.

Every document that moved, every message that got delivered, every connection between departments depended on people like me. We were the nervous system of business, and without us, nothing worked.

Figure 2 – Office at Securit S.A., 1995. Recreated with Nano Banana PRO (2026) based on an original scene described.

None of my time indoors would have happened without Ligia. She had the authority of a leader and the heart of a guardian. From the beginning, she fought to move me away from the street runs and into the office. She saw potential in me that I did not know existed and made sure I had opportunities to grow. Nearly thirty years later, she remained at Securit as head of HR, still doing for others what she had done for me. She showed me that leadership is not about control, but about seeing what someone can become and helping them get there.

Then came a turning point. One afternoon, the accounting department received a computer. Not new to the world, but new to us: a DOS-based machine with a black screen and glowing green text. My colleague approached it with caution, manual in hand, typing with two fingers and cursing every time an error appeared. Weeks later, after gaining some confidence, he asked if I wanted to try. His offer was simple: he would teach me the basics of accounting if I took charge of the computer. Whatever his reason, I only cared about the second part of the deal.

That informal agreement changed things quickly. I started exploring Lotus 1-2-3, the spreadsheet program that ruled offices before Excel took over, building templates and automating calculations that had once consumed my colleague's entire week. A few years later, when Windows spread and Excel became the new standard, I was already prepared to make the switch.

The lesson was one I would see repeated throughout my career. Technology does not simply replace people. It transforms them. My colleague kept his job, but now he could focus on analyzing patterns instead of punching numbers. He moved from calculating to interpreting. Relevance went to those who moved, not those who waited.

By 1996, another opportunity arrived. A machine running Windows. The screen lit up, the startup chime echoed, and something changed in how I saw my own future. At the cost of half my monthly salary, I enrolled at Eurodata, one of the first computer schools in Brazil, and earned my first certificate. I did not know where it would lead, but I knew the direction had changed.

A new tool arrives. The established professionals resist. The curious ones lean in. The balance of value moves. That pattern has repeated itself in every industry and every era, from spreadsheets replacing ledger books to email replacing interoffice mail to smartphones putting the internet in every pocket. Each time, the professionals who thrived were not the most technically gifted. They were the ones who recognized the change early and invested in understanding it before they were forced to. The tool changes. The principle does not.

This is a moment every professional faces at least once: when a new tool arrives, and you have the choice to see it as a threat to your competence or as a bridge to a more valuable way of working.

DuPont: Discovering Technology at Work

By late 1997, I had joined DuPont Brasil as an intern in procurement. On my first day, I was asked to choose an area to work in; options were Sales, R&D, and Procurement. Procurement won simply because it sounded cool, buying things, with no real understanding of how much that choice would shape the next twenty-five years. The job was to migrate printed purchase orders from an IBM Group system into SAP, which DuPont had just begun implementing.

The work itself was repetitive, but the people made all the difference. The senior leaders and other buyers made a conscious effort to include me in every internal and external event from the first week. Waking up each morning with genuine enthusiasm for the day ahead was something entirely new. It was the kind of human environment that makes even the hardest tasks feel worthwhile.

The inefficiencies built into the daily routine became an opportunity. Every buyer started their morning the same way: manually fetching the latest exchange rates to calculate the impact on their own contracts, each one running the same process independently to arrive at the same information. It was duplicated effort dressed up as routine. So, I took that task off their plates entirely, pulling the rates myself each morning, compiling everything into a single report, and distributing it to all buyers through Lotus Notes. One person doing it once, instead of everyone doing it separately every day.

The ABC curve was a different kind of pain. Every month, each buyer I supported had to produce a spend analysis by navigating those green-and-white printed reports by hand, cross-referencing pages to make sense of where the money was going. It was slow, error-prone, and consumed time that could have been spent on actual procurement work. I took over that process too, extracting the data onto floppy disks, building the analysis myself, and circulating the finished reports directly to each buyer. What had been a monthly ordeal became something that simply arrived, ready to use.

For 1997, that was as close to automation as the environment allowed. Not sophisticated by any modern standard, but it replaced duplicated, paper-dependent effort with something consistent, centralized, and digital.

When the migration of over 2,000 purchase orders was complete, privileged access as a key SAP user followed. The role was no longer about processing data. It was about shaping how the company worked. Technology had stopped being just a task and had become a way to contribute at a level well beyond the job title. That early instinct to look at a broken process and find a simpler way, even with the most basic tools available, turned out to be a habit worth keeping.

Figure 3 - DuPont Office, São Paulo, 1998. Recreated with Nano Banana PRO (2026) based on the original scene described. One curiosity about this AI-created photo is that, with a simple prompt on Nano Banana, the AI recreated a photo that is very close to what the office looked like in the 90's.

Building Order from Chaos in the Chemical Industry

At nineteen, joining Laporte Chemicals as a junior buyer felt like the first job with real responsibility. At DuPont, the work had been serious, but there was always the shelter of the trainee title. At Laporte, that shelter was gone. Orders had to go out, suppliers had to confirm deliveries, materials had to arrive, and problems had to be solved before they reached the plant.

The company was in that in-between era of business technology. No longer fully analog, but far from digital as we understand it today. Desktop computers, spreadsheets, email, and network folders existed, yet almost nothing was integrated. Procurement still depended mostly on phone calls, paper trails, and personal follow-up. The full picture did not live in any system. It had to be built manually, one order, one conversation at a time.

In practice, that made the work feel messy and exhausting. A simple sourcing process that today might take minutes within a single platform had to be stitched together across multiple tools and required significant human effort. But for someone paying attention, that kind of environment had an upside. The gaps were obvious. The inefficiencies were impossible to miss. And wherever inefficiency is visible, there is room for someone willing to do something about it.

My mentor, Marcos, introduced me to Microsoft Access, and together we decided to bring some order to the way we handled purchase orders and supplier follow-ups. Neither of us was a programmer. Access at the time allowed non-technical users to build functional databases through forms and fields, not code, and that was exactly what we needed. Over the next year, we built a simple but effective tool that centralized the entire procurement process.

It tracked purchase orders, logged confirmations, monitored deliveries, set reminders for follow-ups, and generated basic supplier performance reports. Before that, most of the day was spent chasing information, digging through folders, flipping through copies of old purchase orders, or calling suppliers just to find out if something had shipped.

The new system changed that. For the first time, a single screen showed what was pending, delayed, or complete. It brought order to a process that had always been reactive, freeing up time for negotiation, planning, and problem-solving instead of hunting for information. Not sophisticated by today's standards, but for a small team without access to modern software, it made a real difference.

That experience planted one of the earliest lessons that still defines how this work gets approached. The real purpose of technology is not to replace people, but to remove the noise that keeps them from doing their best work.

When Laporte was acquired by the German group Degussa, now known as Evonik Industries, the company launched a full SAP ERP implementation. Having built and run the Access system, the transition into the key user role for procurement felt like a natural continuation. That role meant mapping processes, testing modules, logging defects, keeping master data clean, writing quick guides, and running training sessions. Many colleagues were less comfortable with new systems, so the focus stayed on simple steps, clear examples, and making myself available whenever something broke down.

Following the SAP go-live, a promotion to lead buyer followed. The same drive to improve how things worked carried over, with attention on clearer contracts, consistent processes, and better visibility. Not every change landed smoothly. Moving fast and still learning how long-standing habits shaped decisions meant adjusting the approach, spending more time building alignment, and learning to pair results with trust. That period made one thing clear: technology can raise the bar, but people make the change stick. Being effective is not only about systems. It is about timing, buy-in, and helping others succeed with the tools.

That is also the thread running through every chapter of this book. The tools change. Access gave way to SAP, SAP gave way to more sophisticated platforms, and now AI is reshaping what is possible again. But the underlying act, a non-technical professional learning a new tool well enough to improve how work gets done, has never changed. That is not a technical skill. It is a human one.

Unilever: A Bigger Stage, A Harder Start

In 2005, I moved from Evonik to Unilever, stepping into a much larger and more complex organization. The cultural adjustment was immediate and significant. At Evonik, decisions had been quick and direct, in the German tradition. At Unilever, everything operated on a global scale, with layers of structure and processes that revealed a complexity I had no idea existed. I was hired as an Associate Procurement Manager for IT, with no prior background in this area. A few days before I started, the manager who hired me and to whom I would report suffered a serious accident and ended up in early retirement.

On my first day, I found myself running the department alone, responsible for a multi-million dollar spend and high-stakes relationships with high-caliber, multi-billion-dollar suppliers, with just one assistant to help. Suddenly, I was representing Unilever Brazil in global meetings, trying to keep up in a language I did not fully dominate, and reporting numbers that senior leaders relied on. It was intimidating, but it forced me to adapt quickly.

That experience became one of the biggest turning points of my career. Without direct supervision, I had both the freedom and the pressure to shape the department on my own terms. I focused on efficiency and transparency, tracking what was being spent, where, and why. One of my first initiatives was identifying and canceling unused software licenses, which brought significant savings. I also developed a regional purchasing procedure that standardized IT procurement across Latin America, giving buyers in different countries a single framework to work from. All of that with the help of Excel spreadsheets, which were now the dominating tool in the offices.

Those years were intense. I learned that adaptation is not just about learning new skills; it is about moving forward even when the conditions are not ideal. The instinct to move before things feel certain, the same one I learned as a kid in Guarulhos watching cars and as a young office boy navigating the streets of São Paulo, became a standard I would rely on years later. It taught me that waiting for certainty is a losing strategy.

The world moves, and you either move with it or get left behind. That lesson, learned through necessity, is what prepared me for the moment when a machine would challenge everything I thought I knew about my career. By 2008, the intensity of those early Unilever years had earned me a promotion. But what felt like progress on paper would soon reveal itself as the beginning of something very different. The next chapter picks up there, in a role that looked like success from the outside but slowly became the longest period of stagnation in my career.

2

THE HIDDEN PLATEAU

There is a phase in most careers that nobody warns you about, one that does not look like failure. It looks like stability, competence, and a well-earned routine. But underneath the steady performance reviews and the familiar rhythms of the job, something has stopped without you noticing that you are no longer growing. I call this the Hidden Plateau.

Research on career plateaus has consistently shown that prolonged stagnation erodes motivation and engagement even when employees remain technically competent, because the absence of challenge slowly disconnects people from the work itself. What makes the plateau so common is that it does not feel like a problem. It feels like having things under control.

If you have been in your role for several years, you probably know what I mean. The meetings repeat themselves. The deliverables look the same as they did two years ago. You can map your week on a Monday morning and be right about almost all of it by Friday. You are performing well, and that performance has become so automatic that it no longer requires the kind of effort that once made the work interesting. That gap between competence and challenge is where the plateau lives, and most people do not see it until something forces them to look.

This chapter is about learning to see it on your own terms, before it becomes a liability. It starts with my own experience, but the pattern applies to anyone who has stayed in a role long enough for familiarity to replace growth.

The Observer's Paradox

In 2008, I moved from IT procurement to food procurement at Unilever in São Paulo. By all accounts, it was a significant promotion that would offer more visibility and place me at the forefront of key innovations.

My role in IT procurement had been focused on negotiating for enabling technologies like software licenses, hardware, and services. That work kept the business running, but it was indirect spend, costs that supported operations rather than flowing into the products we sold. The transition to food procurement brought me face-to-face with the business's primary inputs. I went from dealing with software to developing strategies for commodities like soybean oil, sugar, and other raw materials.

The crucial difference in impact became apparent very quickly. In my previous role, a successful negotiation might save the company a respectable amount on a software contract. In my new role, a fractional change in the price of a core commodity could affect the company's P&L (Profit and Loss) by millions. This wasn't an indirect expense; this was a direct lever on our Cost of Goods Sold and, therefore, our profitability.

That experience was my first real lesson in identifying an organization's "sweet spot": the areas where decisions have the most direct and significant leverage on financial results. I had moved from a vital support function into the company's economic engine room. The "higher skies" I had envisioned were less about a new title and more about gaining a new, clearer perspective on how a business truly operates and creates value. But the reality soon set in with routine. The role rewarded process compliance more than creativity. I was competent, but I wasn't growing.

At the same time, I had a front-row seat to genuine transformation, just not in my own work. In IT, I had worked with companies like IBM and HP as they reinvented themselves. IBM was repositioning its business model from hardware toward services. HP was expanding its business into mass retail sales of home-use laptops. Microsoft was exploring the cloud. SAP was pushing into business process outsourcing. These suppliers were rewriting their futures, while I kept recycling spreadsheets and forecasts, the world outside was being rewritten.

In 2007, Steve Jobs introduced the iPhone, and by 2008, the App Store had opened, creating an entirely new economy overnight. Companies like Uber and Airbnb were being born, built on platforms that did not exist a year earlier. The suppliers I worked with were racing to reinvent themselves, not because they wanted to, but because they had to. I was watching the transformation from the sidelines, fascinated but passive, not yet understanding that I would soon need to make the same choice.

There was an obsession in Procurement with forecasting exercises, which entailed presenting procurement results in numbers to leadership and predicting where prices and their consequent impacts would lead the business. I hated that because of the operational level, the level of detail involved, and, most importantly, how ineffective that process was. There was no technology to replace the then ultra-complex Excel sheets with data we could not trust, and it would have little to no practical impact on the business.

Switzerland: Settled but Going Nowhere

In 2010, after getting married and a long two-year preparation period, I finally landed a position at Unilever's global procurement hub in Schaffhausen, Switzerland. The office overlooked the Rhine Falls, one of Europe's most beautiful landmarks. This was no ordinary regional office; the Swiss hub was the strategic center of the entire global supply chain function. This was where Unilever concentrated its brightest minds, pulling in top talent and experts from around the World to shape the company's entire global supply chain strategy. From the outside, it looked like the dream step in a global career. The reality was different.

Moving from Brazil to Europe was a complete reset. Except for my wife, I had no close family nearby, no friends, no cultural familiarity. Starting over from ground zero was isolating.

Professionally, the picture wasn't much brighter. In Brazil, I had made decisions. In Switzerland, I was one of several contributors to a slow, layered approval matrix. My scope expanded, but my influence shrank. Projects crawled through committees, and even small changes took months. That was not the organization's "sweet spot" I mentioned earlier.

The emotional weight built gradually. Sunday nights became heavy. For years, Mondays had meant possibility. In Switzerland, Mondays were predictable. If I could map my week in advance down to the calls and outcomes, I wasn't growing; I was stagnating.

The pressure started showing up physically. I had persistent, full-body aches that doctors could not pin to anything specific. Over time, it became hard to ignore the possibility that this was stress, and that the bigger issue was how disconnected I felt from my work and from myself. Everything looked fine from the outside, but the experience of it was heavier and more exhausting than I admitted.

That period taught me something I would only understand years later: adaptation is not just about learning new skills. It is about staying connected to people, to purpose, and to the sense that your work still stretches you. Without that, even success starts to feel like decline. I later realized that technology could automate tasks, analyze data, and predict outcomes, but not create those connections. Every algorithm depends on human behavior, such as judgment, empathy, and intention. As AI takes over more of what we do, what will keep us relevant is how we relate, not just how we perform.

This is how mid-career success becomes an obstacle. The better you get at what you do, the easier it is to stop doing the things that made you good in the first place. The praise continues, but the challenges do not.

The Procurement Repository: An Idea Before Its Time

Out of that frustration, I tried to create something new. My idea was one of a Procurement Repository, a single hub where all knowledge from buyers across the company could be stored and shared: pricing, supplier backgrounds, volumes, negotiation records, savings, and agreements.

The point was simple: when someone took over a role in Procurement, they wouldn't have to start from scratch. They would inherit a complete, organized record of what had been done before. The company would stop losing institutional memory whenever a buyer left their position.

After weeks of work trying to understand the processes and how such a tool would improve the operations, I pitched the idea to my leadership with conviction, certain that it could reshape how we worked. But in 2011, the infrastructure wasn't ready. Cloud platforms were immature, and data integration was limited. The response was a polite rejection: "Not feasible."

Innovation had always moved me, and now there seemed to be no space for it. Years later, when Unilever launched something very similar, I realized I had been right about the direction but wrong about the timing early on. At the time, though, it felt like a door slammed shut in my face.

That same year, IBM's Watson defeated human champions on Jeopardy, demonstrating that machines could process natural language and retrieve knowledge at superhuman speed. The technology I was imagining, a centralized knowledge hub, was already being built elsewhere. I just did not know where to look, and my organization did not know how to see it. We were close enough to sense where things were going, but the distance between that world and ours was still measured in years.

IBM Watson is an artificial-intelligence computer system developed by IBM that can answer questions posed in natural language; it was originally built to compete on the quiz show Jeopardy!, a long-running American TV game show where contestants respond in the form of a question to general-knowledge clues, and in 2011, it beat two of the show's greatest champions.

Looking back, that experience mirrors where we stand with AI today. The technology is powerful, but the world around it is not yet fully ready. We still face limits in infrastructure, energy, data quality, and basic understanding of how these systems truly work. We are once again in a moment where possibility races ahead of readiness. The difference is that now, we can see what's coming. Those who learn to read the signals, prepare early, and adapt as the world catches up will be the ones leading when the timing finally aligns.

The Intern's Approach

The part of my job that still gave me meaning was mentoring. Even when I felt detached from the work itself, I made time for interns and analysts. Maybe it was instinct, or maybe I was trying to hold on to the one part of the job that still felt human.

Maria, an intern from Mexico, stood out from the start with a mix of curiosity and courage that most people lose once they have been in corporate life long enough. One afternoon, she questioned our supplier evaluation model, a tool so familiar that I had stopped thinking about it years ago. Her question was simple: "Why do we assume these are the right criteria?" It should have been a quick clarification, the kind you answer politely and move on. Instead, it showed something deeper, yet simple.

We spent two hours dissecting the logic behind a process I had long accepted as unchangeable. She pushed, and I found myself defending methods I hadn't chosen, only inherited. By the end, we hadn't solved much, but the message was clear. For her, it was just a lively debate, another day of learning.

I realized how far I had drifted from that beginner's mindset, the one that asks "why" before "how." Watching Maria think out loud reminded me of who I used to be at her age: curious, restless, full of ideas I thought could change things. Somewhere along the way, that energy had been replaced by efficiency and routine. Coaching her was rewarding, but it left me uneasy, aware that I wasn't pushing myself the way I once did. I was helping someone accelerate her growth while I stayed parked in place, recycling insights that no longer challenged me.

That afternoon stayed with me because it wasn't really about a spreadsheet or a model. It was about seeing myself through someone else's eyes and realizing how easily experience can turn into autopilot. Maria had reminded me that the moment you stop questioning your own work, you stop growing, and I could feel that truth following me back to my desk long after our conversation ended.

The Illusion Created by Efficiency

Even at the global hub, there was constant talk about efficiency and transformation, yet very little of it turned into real change. We had new systems, dashboards, and tools that looked modern, but they mostly helped us run the same old processes a little faster. It gave everyone the sense that progress was happening when, in truth, we were just polishing what already existed.

I became good at it, too. I built complex spreadsheets, automated reports, and created shortcuts that made the team look productive. But I never stopped to ask the bigger question: Were we improving the work, or simply maintaining it in a more sophisticated way? It was the same pattern I had noticed during that conversation with Maria: the pull toward optimization when reinvention is what is needed. Looking back, I am certain this is not just my story about what I was observing in my job.

It is more common than most people imagine inside large corporations. People talk about innovation while holding on to what feels safe. Tools get smarter, but thinking stays the same. What actually holds organizations back is not inefficiency but the belief that incremental improvements are the same as meaningful change, a belief that keeps both the organization and the people within it busy without moving forward.

Years later, when I saw companies racing to implement AI systems with the same mindset, I recognized the pattern. The tools had changed, but the illusion was the same.

The Acceleration I Missed

In 2012, deep learning researchers at the University of Toronto achieved a breakthrough that would reshape the entire field. AlexNet, a neural network, won the ImageNet competition by a margin so large it stunned the AI research community. It was the moment machine learning crossed from academic theory into practical inevitability. While that was happening, I was in meetings on supplier terms and contract renewals and wasn't aware.

I did not read about it at the time. I was not looking in that direction, and nothing in my professional environment pointed me there. Years later, when I finally understood what had happened that year, the realization hit hard: I had been fully absorbed in the day-to-day during one of the most important turning points in the history of technology, treating each day as a continuation of the one before it, while the tools that would eventually force a complete reinvention of how I worked were being built in labs I had never heard of.

A Hard-to-make Decision

That same year, the question of what came next became impossible to set aside. One morning, sitting in my apartment looking out at what was objectively a remarkable view, something was missing from the picture entirely. The surroundings said one thing, but living in them said another. Staying meant a global title, a predictable career, and the kind of safety that looks convincing from the outside. Leaving meant risk, uncertainty, and starting again. Yet the greater risk, as I kept coming back to it, was staying exactly where and how I was.

That day, I decided to resign. To many people, it looked reckless. I had worked nearly two decades to get there, climbing from a modest start in Brazil to a global position in Switzerland. One director even pulled me aside and said I was out of my mind for walking away from that role, reminding me that thousands would give anything to be in my position. It was a dream most people from where I came from never reached. The decision was not just a career move: in many ways, it meant turning my back on everything I had fought to prove.

Leaving was not easy to justify to myself either. The early years had been hard, learning a new language, finding a way to belong in environments that were not built for someone with my background, building a life piece by piece from very little. All of that had produced what I was now walking away from. Still, I knew that staying would mean slowly becoming a version of myself I no longer recognized.

So, I chose movement over safety. Starting over was a better option than pretending that security was the same as purpose.

The Sustainability Spark

In 2010, Paul Polman, then Unilever's CEO, launched the Sustainable Living Plan. The commitment to double growth while cutting the company's environmental footprint in half was unlike anything I had seen from a major corporation. Many viewed it as a branding move. I did not. To me, it was a signal that the rules of business could be rewritten. It wasn't just inspiring, it was defining.

That vision landed at a moment when I was restless, searching for meaning in work that had started to feel mechanical. Polman's plan showed me that purpose and profit could coexist, that large-scale systems could be transformed from within. It reshaped how I saw my role in the corporate world. I began to think less like a buyer and more like a builder of systems, someone who could connect sustainability, innovation, and human behavior in new ways.

What started as inspiration soon became direction. I poured my evenings into sketching what I called the "Supermarket of the Future," a model built around refill stations, returnable packaging, and circular design. It was a bold idea, and looking back, a bit ahead of its time. But it gave me a sense of momentum I hadn't felt in years. That idea became my ticket into an MIT accelerator, where I learned how to translate vision into models and strategy.

Around that same time, other innovators with whom I had the pleasure to connect were pursuing similar ideas. In Chile, José Manuel Möller founded Algramo, developing refill systems and reusable packaging to reduce waste and expand access to household products. In the United States, Tom Szaky's TerraCycle had already helped popularize new ways to collect hard-to-recycle waste and turn it into reusable materials and products. Seeing models like Algramo and TerraCycle reinforced my sense that circular systems were not just idealistic concepts.

Our paths eventually crossed. I spoke with both founders as we explored potential synergies between our projects. We shared a sense of urgency and a belief that waste was a design problem, not an inevitability. Their courage and experimentation fueled my own conviction that the Supermarket of the Future wasn't a dream, but part of a growing movement.

But despite early encouragement from mentors at the MIT accelerator, scaling the idea proved nearly impossible. Supply chains were rigid, infrastructure was immature, and consumer behavior was not yet ready for refill-based shopping. In hindsight, I was right about the direction, but many years too early on the timing. Still, those conversations validated that sustainability could be both practical and transformative, and that my instincts aligned with a new generation of purpose-driven entrepreneurs rewriting the future of consumption.

A Pivot to Resilience: Education and Entrepreneurship

After the entrepreneurial failure of the "Supermarket of the Future" project, I used my time at the MIT accelerator to immediately pivot my focus back to formal learning. I then enrolled in the Open University for a degree in Environmental Sciences, which served as a crucial bridge between my original career and my new, sustainability-driven purpose.

This renewed focus on core knowledge coincided with a complete change of pace. I needed stability and income, so, moved by a genuine passion, I started boarding dogs in my house through an app called Rover, which quickly grew into Royal Barks, a boutique daycare for dogs with its own location. Clients became friends, and their trust kept me going. That was a powerful demonstration of how human connections can fulfill and bring meaning to any work. Later, I launched Kindmates, a human-grade cooked and freshly frozen dog food company. In its first year, the business gained significant traction: our loyal customer base continued to double quarterly, we achieved a near-perfect retention rate, and we secured a compelling offer to pitch our products to the national buyer at Whole Foods Market. A small operation quickly went from a passion project to a credible, scalable venture.

The Big Leap of Faith, Collapse, and Return to Corporate Life

Crucially, I designed Kindmates around a core sustainability principle. To avoid the wasteful cardboard boxes and excess insulation materials used by today's competitors like The Farmer's Dog and Ollie, I invested in high-performance, reusable insulated shipping containers from Liviri, a subsidiary of Otterbox, the makers of high-performance iPhone cases. Those boxes only required a few ice packs to keep food frozen and were designed for multiple cycles of use, dramatically reducing waste. Each container was shipped to clients with a return label, and customers left the boxes on their porches for convenient pickup and return. Although this commitment to reusable packaging significantly reduced my profit margins, it was a non-negotiable part of my strategy to minimize environmental impact and uphold our sustainability values.

The business was at full steam and had big plans for 2020. Then COVID hit. Over the next few weeks, the ghost kitchens we used to produce Kindmates' products began to close, and the supply chain rapidly collapsed. Delivery services were operating at capacity, resulting in major nationwide delays. After several smaller logistical setbacks, the system failed completely: a carrier missed the guaranteed delivery to my largest single client, who spent more than $1,000 a month on food for his three large dogs.

Knowing I couldn't risk further delay for this customer and that I'd already lost the cost of the first order, I took an extreme measure: I prepared a replacement order and personally drove 11 hours from Florida to North Carolina to ensure a timely delivery of the fresh food.

Despite this extraordinary personal sacrifice, the client, an influential CEO, was intransigent over the overall delay and reacted with hostility, threatening to undermine my business and refusing to return the valuable, custom-made container. His intransigence meant I had lost the value of the original order, the cost of the replacement order, and the twenty-two-hour road trip I had taken. The symbology of losing that important client, coupled with the momentary stress, supply chain issues, and a loss of financial oxygen, forced me to close the business in early 2021.

Lessons from the Plateau

That period, from the promotion to the Foods Procurement team in 2008 through the collapse of Kindmates in 2021, held real growth and genuine momentum, but it also showed me that stagnation is not neutral. It does not announce itself but builds gradually, one routine meeting at a time, until repetition starts passing for expertise. What depletes a career is not failure but predictability. The praise keeps coming, the performance reviews stay positive, and somewhere along the way, you stop asking whether any of it still matters.

Looking back, those years were among the most important lessons of my career. They taught me that safety can be as dangerous as failure, because it numbs the desire to evolve. I began to understand that growth is not something you chase only when things are falling apart; it is something you must protect, even when life looks stable.

Switzerland taught me that professional stability can hide personal decline. When growth stops, even success starts to feel hollow. The entrepreneurial years taught me something different: that even bold moves can fail if the timing and the infrastructure are not ready. Both lessons pointed in the same direction. Standing still is not safe; it is surrender.

If you have spent years in the same role, function, or organization and recognize the symptoms described above, such as Sunday night heaviness, the ability to predict your own week, and praise that no longer connects to growth, know that this is a structural feature of careers that reward consistency over reinvention.

Research on professional development consistently shows that expertise plateaus when the challenge disappears. The hidden plateau is not something that happens to a few unlucky individuals. It happens to every professional who stays in a role long enough for routine to replace challenge. Recognizing it is the first real sign of awareness.

But failure, as painful as it was, turned out to be a more honest teacher than comfort ever had been. And it was from that low point, depressed, financially stretched, and uncertain about what came next, that the next chapter of my career began.

Entrepreneurial failure teaches a lesson that corporate success often hides: your resilience is more durable than your plans, and that starting over does not erase what you have learned. It adds to it.

3

STARTING AGAIN, STARTING DIFFERENT

Back to the Future

In early 2021, depressed, directionless, and with my pride badly hurt, I had lost nearly everything. It was at that low point that I decided to return to corporate life. That decision eventually led me back to Unilever USA, where I could apply the difficult lessons learned about supply chains and resilience during my entrepreneurial years. I returned to a familiar procurement department, but I was different. Not just older, but changed in how I saw work, risk, and what was worth my attention. Supply chain disruptions were not just crises: they were pattern problems waiting to be solved. Vendor relationships were not just transactions: they were networks that technology could strengthen.

Shortly after returning, while working full-time at Unilever, I began a master's program in Sustainable Business at the University of Wisconsin. It was an intense period, but it connected two things I had kept largely separate: the academic frameworks of systems thinking and the daily realities of running supply chains under pressure. That adjustment in thinking was already permeating how I worked. I was becoming faster at identifying patterns, quicker to question delayed information, and more willing to act on early signals rather than wait for consensus.

That became clear when a severe drought in Canada disrupted mustard seed crops, putting Unilever's condiment lines at risk, a product that sat directly in my portfolio. For most, it looked like an inevitable supply crisis. To me, it was a reason to move early. Instead of relying on delayed reports and expert opinions, I flew to Calgary, met with farmers and brokers, and negotiated supply agreements directly. By the time the shortage made headlines, our production was secure.

It was not luck but adaptation, a combination of experience, intuition, and speed that had been developing long before that drought made headlines. Resilience is the product of awareness and well-built relationships, developed well in advance of a crisis, not something that emerges in the middle of one.

As the master's program progressed, I began experimenting with AI tools to manage the growing demands of studying while working full-time. Perplexity. ai became my shortcut for research, helping me cut through endless academic papers, while Google's NotebookLM helped me organize notes and turn them into coherent insights. These tools did not just make learning easier; they changed how I thought and worked. Soon I was showing classmates how to use them and suggesting to professors that they set clearer guidelines for AI-assisted research. It was an early sign of how technology could change the way we learn and make decisions, and a preview of what was coming next in my professional life.

AI for Non-Techies

At Unilever, I began exploring the AI resources the company was making available. Like many large organizations, Unilever approached the technology with caution, building boundaries around sensitive data while still giving employees room to experiment with approved tools. I joined every AI-related initiative I could find. When Unilever launched an AI Champions group, I was among the first to sign up. I attended internal workshops, participated in pilot programs, and volunteered for any AI-related project that came up.

What became clear early on was that most colleagues were curious about AI but were not using it. The gap between interest and action was wider than anyone was acknowledging. So, I started running what I called AI for Non-Techies workshops, beginning as informal one-on-one sessions with anyone who expressed interest. I would show practical applications, demonstrate real tools, and help people see how AI could make their specific jobs easier and more interesting. Those sessions grew into small-group workshops and eventually into larger presentations, reaching hundreds of people across different functions and teams.

Each session was tailored to the group's specific challenges and use cases. The response was consistently positive, yet the same pattern kept emerging. People were amazed by what AI could do in the room, but few carried it into their daily work afterward.

Working With What You Have

One thing I learned about AI adoption in corporate environments is that there is a significant distance between enthusiasm and implementation, and that these are very different things.

The security boundaries that protect company data also limit AI functionality. Even within Microsoft's Copilot environment, which was already useful for meeting summaries, email responses, and presentation creation, certain features were restricted. Employees could not always experience what the tools were fully capable of, which made it harder to reach the moment of genuine conviction that turns curiosity into consistent use. It was a real limitation, but understandable.

Organizations of that size and complexity cannot afford to have sensitive data accidentally shared with external AI systems. The challenge is creating enough functionality within secure environments to demonstrate real value. Despite those boundaries, what was available was already producing results. Copilot's ability to generate meeting notes with action items, help draft professional email responses, and support presentation development was already producing real-time savings for those who used it consistently.

The lesson was straightforward: do not wait for perfect tools to be approved. Start with whatever is available and build proficiency with those. When more advanced capabilities eventually open, the advantage goes to those who are already fluent, not those who are still getting started.

The Shark Tank Moment

The real breakthrough came when Unilever launched an internal competition modeled on the Shark Tank format, giving employees the chance to present bold initiatives that addressed long-standing problems but had never had the time or resources to be explored.

For me, the starting point was something I had been thinking about for a while. In large organizations, procurement teams often work within review cycles that create a natural lag between what is happening in the market and what the business can do about it. By the time a trend surfaces in a quarterly report, the moment to act on it has frequently already passed. The question I kept coming back to was whether technology could close that gap, giving buyers real-time visibility into market movements and their financial implications so that procurement could get ahead of developments rather than respond to them after the fact.

That idea became my pitch. A platform that connected commodity market data to financial impact modeling, designed not as a reporting tool but as an early signal system that procurement teams could act on before the situation required escalation. It was the kind of tool that did not yet exist inside the organization, and the competition felt like the right moment to make the case for building it.

The preparation was not a solo effort. I built a cross-functional team of six, drawing people from procurement, finance, commodity risk management, and IT. Finance helped model how savings or losses would affect business results. The risk management group guided the logic of commodity allocations. IT assessed which technologies could support the vision. Many working sessions shaped the project from a rough concept into something that looked like it could work.

When the day arrived, the format mirrored the television version: live pitches to a panel of senior executives from across different departments, in front of the entire global procurement community. Forty-five minutes to make the case. I did not frame the proposal as a dashboard or a reporting tool. I framed it as an AI product, a forward-looking application that did not yet exist but could change how decisions were made. That framing mattered. It positioned the idea not as an incremental upgrade but as a fundamentally different way of working.

When the panel announced my project as the winner, what it gave me was not a promotion or a pay raise, but something more useful at that stage: credibility. For the first time within the organization, people could see me not only as a buyer but as someone thinking about the future of how the work got done.

Mindset Transformation

Winning the competition opened a door that had not existed before. The transition into an AI Product Owner role brought a different kind of work: coordinating across multiple products and stakeholders. AI became the tool that made that coordination manageable. I created what I called cockpits, Projects within ChatGPT that acted as dynamic knowledge systems for each product I managed. They were not file repositories; they were working environments that could retrieve context, surface relevant information, and answer questions in real time. As of the end of 2025, these capabilities are still developing, with ChatGPT Projects among the more advanced examples, and similar approaches beginning to emerge in tools such as Perplexity AI and Claude.

The most significant change was not in the tools themselves but in how work felt. AI did not just make things faster; it made it possible to be more deliberate about how time and attention were spent. Repetitive tasks that had previously consumed hours were handled differently. Formatting, tracking, and synthesizing information across systems became less burdensome, freeing up more capacity for work that required judgment.

Through that experience, something became clear that I now think of as the compound effect of AI. Most people encountering these tools for the first time see them as a faster calculator or a smarter search engine. But when AI becomes part of the daily workflow, it does not just improve efficiency at existing tasks; it gradually expands what is possible to take on altogether.

The Evangelist's Dilemma

The success of the internal workshops led to opportunities outside Unilever. What kept repeating across organizations was the same pattern: high curiosity, low adoption. People were genuinely interested in AI's potential but struggled to move from awareness to action. The gap was almost never about technical knowledge but about imagination. They could not picture how these tools would fit into their specific daily work. So, the workshops evolved. Less time on what AI could do in general, more time on what it could do for the specific person sitting in the room. Walking people through use cases they could try the next morning worked better than any demonstration of general capability.

That adjustment in approach changed the workshops' output. People adopt tools that address their specific problems, not tools that seem impressive in the abstract. Learning to start with the audience's challenges rather than the technology's features was the adjustment that made the difference, and it also changed how I thought about the role of an early adopter.

Being ahead of a curve brings a responsibility that is easy to underestimate because people who have already made a transition tend to forget what it was like before they understood the tools. That process taught me that real progress comes from staying patient and meeting people where they are, rather than pulling them toward where you already are.

Finding a Tribe

Until the Champions network, most of my AI learning had been a solitary process: late nights with ChatGPT, podcasts during commutes, books laid on the desk. The network changed that. Being around people working through similar questions, how to experiment within policy constraints, how to translate abstract potential into something practical, and how to bring colleagues along without losing them made the whole process less isolating. Learning alongside others who were genuinely curious but not yet certain produced a different kind of progress than learning alone. Seeing how different people approached the same problems from different angles made the thinking more effective.

Cultivating Curiosity

Many colleagues reached out privately, asking about tools they could use in their personal lives rather than those approved by the company. That distinction mattered less than it might seem. Getting people at ease with experimentation was the first step, regardless of context, because someone who starts using AI at home to organize their studies, plan a trip, or draft a personal email will eventually bring that confidence into their professional work. The two are less separate than organizations tend to assume.

Running the workshops also changed how I prepared for them. Teaching required something beyond knowing the material; it required listening first. Before each session, short conversations with attendees to understand what they wanted to see. Principles borrowed from adaptive learning, tailoring the content to the audience rather than delivering the same session each time. That approach produced better results in the room and taught me something useful outside of it: the ability to communicate in ways that connected rather than simply informed.

One moment from that period stands out. A colleague who had been openly skeptical attended a session to see what the discussion was about. By the end, he asked for a follow-up conversation. Weeks later, he mentioned that he had begun using AI regularly for his reports and could not quite believe he had waited if he had.

A New Identity

For two decades, my introduction had been the same: a procurement professional. The change into the AI Product Owner role changed that. Not just the title, but the way I was seen and, gradually, the way I saw myself. No longer the procurement professional who knew AI, but the AI professional who knew procurement. That reframing opened conversations and opportunities that had not been available before.

The Shark Tank pitch had worked not because of any gift for presenting. It worked because the approach was repeatable. A real operational problem was identified. A working mockup was built using Lovable over a weekend. The pitch was framed around business impact rather than technical novelty. Something tangible was shown rather than described. The gap between just having an idea and showing a version of it is smaller than most people assume, and it changes how every conversation about that idea goes afterward.

Outside of work, the transition carried its own kind of relief. Friends, family, and acquaintances stopped associating me exclusively with a buyer's career. The label that had defined me for two decades no longer applied in the same way, and that openness felt significant in ways that went beyond the professional.

The first months in the new role were not easy. Leading a team in an area without a formal background meant regular encounters with the edges of what was known. The difference was trusting the colleagues around me and allowing them to fill those gaps. Their contribution anchored the work while the new territory became more familiar. Reinvention does not mean doing everything alone, and that period made that clear in practical terms.

The message for anyone facing a similar crossroads is this: domain knowledge is not a limitation in an AI age; it is the foundation. Whatever field you are in today, that expertise can be translated into AI-driven contributions tomorrow. Reinvention takes what already exists and reorganizes it into something that works in the current environment, nothing erased, everything repurposed.

Learning to work with AI did not end the career as it had been. It changed how work, learning, and relevance were approached altogether. But crossing that threshold also raised a different question. Personal reinvention is one thing; translating it into something others can use is another. That is where the ADAPT method came from.

The next part of this book moves from personal story to practical framework. The same principles that carried the journey through disruption, stagnation, and reinvention can be applied across careers and industries. The goal is not to become a technologist. It is to recognize signals early, set a direction, act before certainty arrives, position what you know, and use technology as a multiplier. Part II is where the experiences become the playbook.

From Memoir to Method

The story in Part I, from the early years in Brazil to the AI Product Owner role, is more than a personal history. It is a pattern. The forces that shaped that career, technological disruption, periods of stagnation, and the recurring need to reinvent are the same forces most professionals are navigating right now.

Part II moves from the why to the how. It begins with the Polymath Approach, the mental model that supports thriving in an environment where AI handles depth at speed, and the advantage goes to those who can connect ideas across domains. From there, it moves into the five steps of the ADAPT® method: Awareness, Direction, Action, Positioning, and Tech Empowerment, each with practical frameworks, real examples, and exercises designed to produce results rather than just reflection.

The platform at adapterslab.ai runs alongside every chapter, extending the exercises into an interactive environment that updates as the tools and best practices evolve. The book provides the framework. The platform keeps it current. Together, they are designed for one purpose: to help you move.

THE ADAPT METHOD

A practical framework for readers to use. Awareness, Direction, Action, Positioning, and Tech Empowerment become step-by-step moves with prompts, exercises, and playbooks. This section distills lessons from the author's story into a repeatable path that does not require a technical background.

4

MASTERING THE POLYMATH APPROACH

For most of my career, I was told the same thing many of you have probably heard: pick one path, stick with it, and build expertise until you become the go-to person in that field. That has been and remains the rule of the game in the corporate world. Specialists were rewarded. People who went deeper in one area were the ones with a clear path to success.

But the game is changing. AI is taking over many of the repetitive, predictable, and even technical tasks that specialists once considered their own domain. What looked like a safe path five or ten years ago is now under pressure. Roles that depend on doing one thing extremely well are the most at risk. What the world now rewards are people who can move across fields, learn quickly, and bring ideas together in ways others do not see. That is the essence of being a polymath.

The term "polymath" comes from the Greek *polymathēs*, meaning "having learned much."

Historically, polymaths were scholars who achieved mastery across multiple disciplines. Think of figures like Leonardo da Vinci, who excelled in art, engineering, anatomy, and physics, or Benjamin Franklin, who moved fluidly between science, politics, writing, and invention. During the Renaissance, this breadth of knowledge was celebrated as the ideal of human potential. However, as industrialization took hold and academic disciplines became more specialized throughout the 19th and 20th centuries, the polymath became rare. Universities created siloed departments, corporations built vertical hierarchies, and the specialist became the model of success. The polymath was increasingly seen as someone who knew a little about many things but mastered nothing.

Today, that perception is being challenged. A polymath is not just someone curious about many things. It is someone who develops depth in more than one field and then combines those fields to create something new. They are not "all over the place." They are building bridges between areas of knowledge. In the age of AI, that ability is becoming one of the most valuable skills you can have.

Living the "Everywhere" Life

For a long time, I worried about how my own path looked to others. I worked in procurement, then tried entrepreneurship in the pet services industry, went back to school to study environmental sciences, spent time as a researcher in the food industry, later immersed myself in AI, and finally became an author. From the outside, these disconnected moves could appear to be a lack of focus. At least a couple of times, I even heard casual comments from colleagues and family that I was "always doing something different" in a tone that sounded derogatory to me. At times, I wondered if I was unfocused, if my career lacked a single narrative.

Only later did I realize those moves were not random; they were the foundation of polymathy. Each move gave me tools I would eventually use when AI entered my world. Procurement taught me analysis and negotiation. Running a dog business taught me entrepreneurship and customer empathy. Sustainability studies taught me systems thinking. Volunteering showed me how small actions can have a big impact. Even personal health transformation through running a marathon taught me discipline and iteration.

Separately, these events looked like detours. Together, they became a portfolio. And when the moment came to step into AI, I didn't start from zero. I brought perspectives and skills that a pure specialist wouldn't. That is the point: variety becomes a strength once you learn to connect it.

Why Polymaths Matter in the AI Age

Research shows why polymathy is so powerful in today's environment:

- **AI automates routine.** The more repetitive and narrowly defined a job is, the easier it is to replace. Polymaths thrive because they can change across domains when one area changes.
- **Big problems cross boundaries.** Climate change, digital transformation, and healthcare: none of these can be solved from a single lens. Polymaths bring multiple perspectives together.
- **Organizations need translators.** AI projects require business leaders, technical experts, and end users to understand each other. Polymaths bridge those gaps.
- **Adaptability beats certainty.** Specialists often feel stuck when their domain moves. Polymaths have already practiced adaptation and can pivot faster.

This idea is echoed in a recent article on *Psychology Today* titled *"Cultivating the Modern Polymath with AI"*, written by psychologist Andrea Quesada, which describes how AI is helping polymaths multiply their impact by supporting fast learning across disciplines. In short, polymaths do not compete with AI. They expand what's possible.

The remainder of this chapter focuses on practical techniques to help you build your own polymath path. These tools are available online at adapterslab.ai/polymath, so you can explore them there and continue directly to the next chapter if you wish.

How to Build a Polymath Mindset

You do not need to master ten fields to start building your polymath's journey. What you need is a way to learn and connect what you already have with what you can have. Here's a simple and practical approach:

1. **Start with an anchor.** Go deep in one field that gives you credibility. For me, that was procurement. If you are a seasoned professional, a bonus.

2. **Add branches.** Explore two or three areas that interest you and push you outside your usual area. These can be hobbies, studies, or side projects. Sustainability, entrepreneurship, and technology became mine.

3. **Look for patterns.** Practice asking: *What can I apply from here to there?* Volunteering at a Farm for a couple of weeks taught me about systems and processes. Fitness taught me about consistency. Entrepreneurship taught me discipline. AI became the field where everything merged.

4. **Use AI as a partner.** Leverage AI tools to accelerate research, test ideas, and organize learning, but keep human synthesis and judgment at the center, meaning you are in control, not the machine.

5. **Tell your story.** Do not hide your variety. Frame it as integration. Employers and clients value people who can connect dots across silos.

The Human Edge

AI is advancing at a pace that feels unstoppable, but the connections it makes across fields come from pattern recognition in data, not from having lived inside those fields. It can analyze, optimize, and generate at remarkable speed, but the intuitive leaps that come from years of experience across multiple domains remain distinctly human work. A machine will not work as a teacher for years, learning how to break down complex ideas for students with different learning styles, and then bring that insight into product design to make software more intuitive for non-technical users.

It will not spend time as a journalist, trained to ask the right questions and verify sources, and then apply that investigative mindset to due diligence in venture capital. It will not learn project management in construction, where delays in one phase cascade through the entire timeline, and then use that understanding to redesign how a hospital coordinates patient care across departments.

These connections require more than intelligence. They require context, empathy, pattern recognition shaped by personal narrative, and the ability to see meaning where others see only data. AI can assist us once we've made those connections. It can scale, refine, and execute them with speed and precision. But the spark, the moment of synthesis that creates something genuinely new, remains exclusively human work. It is our job, the human job, to give AI direction, purpose, and vision. Without us, Humans, AI is a powerful tool with no place to go.

That is why polymathy matters now more than ever. It is not about knowing everything but seeing connections and using them to create solutions that are hard to copy.

Reflection: Your Polymath Map

Many of us already live polymathic lives without naming them as such. The challenge is not gathering more experiences; it is connecting the ones you already have.

Here are some questions to map your polymath path:

- **Anchor:** What's the field you know best today? Example: If you are a nurse, then Healthcare is your field.
- **Branches:** What two or three other areas have you explored, through work, study, or even hobbies?
- **Connections:** Where do you see overlaps between them?
- **Application:** How could those overlaps solve a problem you face now?
- **Next Step:** What's one new area you'd like to explore that could complement your anchor?

Write these answers down. Look at the map you produced. You will begin to see how what once looked like "being everywhere" can be the structure of a polymath career. Check the Resource section below or visit adapterslab. ai to create your digital version of the Polymath Map.

In a world where AI takes the routine, polymaths create the future. The advantage is not in doing more of the same, but in connecting what seems unrelated into something new. That is the mindset worth pursuing.

Resource:

In adapterslab.ai/polymath-map, you have the resources you need to start creating your polymath mapping. You may use the criteria below to guide your mapping journey.

Anchor: Start with one field of deep expertise. This is your foundation and credibility.

How to apply it:

Imagine you are a *marketing* professional. That is your anchor. You understand audiences, messaging, and brand strategy. Instead of abandoning it, you use it as a base to explore other interests. For example, you might study *psychology* to better understand consumer behavior or learn *data analytics* to make campaigns more evidence-driven.

Or imagine you are a *nurse*. Your anchor is patient care, reading subtle cues, managing crises, and communicating under pressure. You can build from that foundation by learning about *digital health tools*, *AI diagnostics*, or *data tracking* for patient outcomes. Those new skills do not replace your nursing expertise; they expand it. You move from caring for patients one-on-one to helping shape how healthcare itself evolves.

Or you are a *teacher* who decides to learn *Vibe coding* to create digital learning tools for your students. The teaching experience stays central; vibe coding simply extends the reach of that expertise. The point is not to start over, but to deepen and extend what you already do well.

Branch - Choose two or three other areas that spark curiosity. These can be professional or personal.

How to apply it:

Let's say you are an HR leader curious about sustainability and AI. You take a short course on ESG (environmental, social, governance) metrics and experiment with AI recruiting tools. Suddenly, you are not just managing talent, you are helping your company align hiring practices with its sustainability goals.

Or if you are a designer who also loves storytelling and psychology, explore how narratives shape user experience. The mix of disciplines becomes your competitive edge. Branching means following interests that strengthen your main skill, not chasing whatever happens to look appealing at the moment.

Connect: Look for patterns and overlaps between fields. Ask: "What can I apply from here to there?"

How to apply it:

A software engineer who practices martial arts might notice parallels between coding and movement. Both require rhythm, structure, and discipline. That insight could inspire better team flow in agile development.

A chef learning sustainability might realize that food waste management principles apply to time management and resource allocation in business.

These connections often happen by accident, but you can train yourself to notice them. When reading, ask: "Where else could this idea fit?" The mind becomes a bridge instead of a box.

Amplify with AI: Use AI tools to accelerate research, organize learning, and test ideas, but keep your human judgment at the center. AI is made for humans, by humans.

How to apply it:

A journalist can use ChatGPT or Perplexity to summarize dozens of reports in minutes, then spend their creative energy analyzing what those reports mean.

An architect might use Midjourney to create and visualize design concepts faster, freeing time to think about cultural or emotional aspects of space, things no algorithm grasps.

The key is to let AI handle the grunt work while you focus on the insights. Let AI handle the grunt work while you handle the interpretation.

Tell the Story: Frame your variety as integration, not distraction. Show how your experiences fit together to solve problems.

How to apply it:

When updating your LinkedIn or interviewing, do not say "I switched careers a few times." Say, "My background in finance and psychology helps me design products that make complex decisions simple for users."

If you have worked in nonprofits, tech, and marketing, connect them through a shared purpose; perhaps you have always been driven by how people adopt change. The story turns from 'I have done too many things' to 'I bring unique combinations of experiences that others do not. Even internally at work, tell micro-stories that show your multidisciplinary thinking, like "Here's how lessons from my last project in X helped me solve Y."

Keep Learning: Polymathy is never finished. Treat every new skill or project as raw material for future connections.

How to apply it:

Set a rhythm: one mini-course, new book, or skill every quarter. Keep a "learning log" where you jot down how each new topic connects with your main field.

For example, a finance manager could start a short podcast or LinkedIn series about leadership and decision-making, learning communication and storytelling skills along the way. A nurse curious about technology might test wearable health devices to understand how patient monitoring is evolving.

The goal is not to change careers overnight, but to explore new spaces that connect back to what you already know. Each small experiment adds a new layer of understanding and expands the range of problems you can solve later.

Think of yourself less as a job title and more as an evolving system of skills. The question worth asking is not 'What do I do?' but 'What can I do next with what I already know?

Becoming a Practicing Polymath (Without Burning Out)

A polymath connects knowledge across a few different areas and uses those intersections to solve problems in smarter ways, without needing to know everything or maintain five careers at once. You do not need to go back to school or master ten disciplines. You need a system for curiosity and a habit of application.

A set of simple practices I've used, and seen others use, to build a polymathic edge looks like this:

Keep an "Adjacent Curiosity" List

Polymaths do not chase every trend they see on Instagram or LinkedIn. They follow threads connected to their current world. Are you in marketing? Explore psychology. Do you work in logistics? Check into systems thinking or urban design. Keep a running list of topics that sit near your zone. Whenever something grabs your attention, write it down, even if you do not have time to explore it yet. My current adjacent curiosities are plant propagation, small-cap investing, and child psychology. These seem very unrelated and random, but in reality, they are distinct fields that complement each other, forming the foundation of a polymath.

Use a 70-20-10 Learning Split

This is less a strict rule and more a helpful rhythm.

- Seventy percent of your learning time goes to deepening your core field; that is your anchor.

- Twenty percent goes into adjacent areas. Think about tools, trends, or methods that influence or touch your main work.
- Ten percent goes into "wild cards." That is where polymath sparks often happen. For me, it was things like dog nutrition, sustainability, and eventually AI.

Set Up an AI-Powered Second Brain

Use tools like Notion, Obsidian, or ChatGPT Projects to capture and connect your ideas. Do not just store articles or notes. Add short reflections in the form of prompts: "How could this apply to my work?" or "What does this remind me of?" Over time, this turns passive reading into active synthesis.

Follow Pattern, Not Platform

When trying a new skill or field, do not just dive into tools; first, learn the basics. Look for *patterns*. What mental models do experts use? What questions do they ask? For example, learning product design is more than mastering the app Figma or UI (User Interface) kits; it is about how designers think through user journeys and tradeoffs. That mindset can apply across industries and sectors. It is all about the process.

Start Small, Then Build

Do not try to master everything at once. Choose one project where you can apply two skills together, say, writing and data analysis, or AI and sustainability. Even something personal, like building a hobby site or automating part of your budget tracking, counts. Polymathy is built in layers. One overlapping project at a time.

Teach, Even Informally

You do not need to become a course creator. But when you explain something to someone, through a LinkedIn post, a short YouTube video, or a workshop for your team, you *connect the dots* in ways that deepen your understanding. Teaching forces synthesis. That is a polymath's instinct.

Keep a Wins Log

Track little moments when your mix of skills gives you an edge. Maybe you solved a problem faster because of a random book you read or a YouTube video you watched. Maybe your fitness hobby helped you lead a better workshop on energy management. Log them. This helps you *see* how your range creates value and builds your confidence to keep going.

Future-Ready Careers for the Polymath Mindset

The world has stopped rewarding specialists who know one thing and do it on repeat. AI is already good at that. The world's asking for connectors, synthesizers, translators, people who can see across boundaries and bring unusual combinations together.

Here are some roles and career paths that are either emerging now or becoming more central and that *reward* polymathic thinking. If you are reading this book from a completely different background, do not worry. These are not new job titles you need to go back to school for. They are *shapes* of work that pull from multiple skill sets. You may already be halfway there.

1. AI Translators and Workflow Designers

Not everyone needs to build AI. But someone needs to *translate* business problems into prompts, tools, and prototypes and help teams implement them. These are people who understand how a process works *and* how to use AI to simplify or improve it.

Example: An HR manager who learns how to build an internal chatbot for onboarding, combining people skills, workflows, and no-code tools.

2. Hybrid Product Owners

Traditional product managers focused on features and roadmaps. The next wave will include *AI-native product owners* who combine business context, UX sense, data fluency, and just enough tech to shape AI-powered tools.

Example: A marketing strategist who pivots into building personalized AI-driven customer experiences, using insights from campaigns, copywriting, and tech testing.

3. Learning Designers for the AI Age

With upskilling becoming the norm, organizations will need people who can design learning experiences that are fast, engaging, tailored to different roles, and powered by AI. These people mix communication, psychology, content creation, and tool fluency.

Example: A teacher or coach who moves into corporate Learning & Development using AI to build adaptive learning journeys.

4. Personal AI Consultants / Side Hustle Engineers

As AI becomes part of everyone's daily workflow, there is a growing role for advisors who help individuals, not companies, build and organize their personal set of tools. This is part coach, part prompt whisperer, part system builder.

Example: A freelancer who helps solopreneurs automate client communication, research, and admin using off-the-shelf tools.

5. Digital Ethicists and Trust Builders

AI is powerful, but it needs oversight structures. Roles that blend human behavior, tech awareness, and policy are gaining ground, especially in large organizations or regulated industries. Example: A compliance officer who grows into an AI governance lead, bringing their legal mindset together with an understanding of model risks and biases.

6. Creators Who Multiply

Writers, designers, researchers, and strategists who learn how to multiply their output using AI will stand out. The value is not just creating; it is creating *at scale* while keeping a human voice or lens. Example: A graphic designer who uses Midjourney, ChatGPT, and Canva AI to run a micro-agency for fast-turn branding projects.

7. Problem-Solving Generalists

As teams flatten and tech gets easier to use, the best people will often be "utility players" who can jump into strategy, operations, experimentation, or customer support and stitch them together. Example: Someone from customer success who learns prompt design, builds simple automations, and becomes the go-to person for AI-enhanced service.

How to Spot the Right Path for You

If you are unsure where you fit, do not start with job titles. Start with your overlaps. Ask yourself:

- What's my anchor skill?
- What are the weird things I've done on the side that taught me something useful?
- Where are other people stuck, and what could I combine to help them?

Most future roles will not come with very clear labels and descriptions. They will come from people like you, who combine what they already know with what's possible now and invent something others did not see coming.

Mini Frameworks for the Polymath-in-Progress

These four tools are not exercises to complete once and file. They are instruments you return to at different stages: when you feel stuck between disciplines, when you are building something new, or when you sense you are capable of more but cannot name what that looks like yet. Use whichever one matches where you are right now.

1. The Anchor + Branch Map

What it is for: *Clarifying your starting position before you move.*

Most people who want to think across disciplines make the same mistake: they abandon what they know in pursuit of what they do not. The Anchor + Branch Map works against that impulse. It shows you that your existing expertise is not a limitation to escape; it is a platform to build from.

Use this when you are feeling unfocused, when a career transition feels overwhelming, or when you want to explore something new but are not sure where to begin.

How to use it:

- **Anchor:** Your current field, role, or most practiced skill. Be specific: not "business" but "vendor negotiation" or "data reporting."
- **Branch 1:** A side interest or experience you have let go quiet. Something you were drawn to before life got in the way.
- **Branch 2:** A tool, topic, or trend pulling at your attention right now. Not what you think you should be learning, but what you keep clicking on.
- **Bridge:** A real project or problem where all three could meet. This is where the polymath lives.

Example:

- **Anchor:** Procurement
- **Branch 1:** Sustainability certifications (studied years ago, never applied professionally)
- **Branch 2:** AI-powered data dashboards

- **Bridge:** Build an internal tool that tracks supplier sustainability scores using AI to flag risk and surface insights automatically

The bridge does not have to be a finished product. It could be a proposal, a proof of concept, or just a question worth investigating. What matters is that it connects what you already are to what you are becoming.

2. The "Borrow and Build" Prompt

What it is for: *Solving problems by stealing intelligently from other fields.*

Every discipline has solved problems that other disciplines have not noticed yet. A polymath's advantage is pattern recognition across those walls. The "Borrow and Build" prompt is a structured way to do what polymaths do naturally: take a mechanism that works somewhere and move it to a place where it has not been tried.

Use this when you are stuck on a problem that feels unsolvable in your current context, when you are designing a new process, or when standard approaches in your field have stopped producing new results.

The prompt:

"I am taking the way [X field] does [Y concept or mechanism] and applying it to [Z context]."

The specificity matters. "X field" should not be vague ("technology" or "business"). It should name a practice: what surgeons do before operations, what screenwriters do in act breaks, what supply chain managers do with buffer stock.

Examples:

- "I am taking the way emergency rooms do triage and applying it to my team's backlog of support requests." Instead of working chronologically or by whoever shouts loudest, you categorize tasks by urgency and impact before touching any of them.

- "I am taking the way product managers run sprints and applying it to my team's learning and development plan." Two-week cycles, a defined goal, and a retrospective. Instead of a vague annual training calendar, your team works toward specific skills in focused bursts.
- "I am taking the way Netflix runs A/B tests and applying it to how our department writes internal reports." Send two versions to different audiences, measure which format drives more action, and standardize based on what works.

The power lies in the habit of asking: who has already solved a version of this? What did they learn that I have not borrowed yet?

3. The Polymath Stack Builder

What it is for: *Designing your toolkit with intention, not accident.*

Professionals tend to accumulate skills reactively: learning what the job requires, what a course offers, and what a manager suggests. The Polymath Toolkit Builder is a way to be deliberate about how your capabilities combine. A toolkit is not just a list of things you can do. It is a theory of the unique value you create when those things work together.

Use this when you are preparing for a new role, building a personal brand, designing a side project, or simply trying to articulate what makes your perspective unusual.

Build your set of AI tools by selecting one from each category:

- **Mindset Skill**: How you approach problems. Systems thinking, design thinking, coaching, storytelling, and first-principles reasoning.
- **Tech Tool**: What you build or analyze with. ChatGPT, Notion, Figma, Airtable, Midjourney, and Power BI.

- **Field Lens**: The domain that shapes your frame of reference. Education, psychology, sustainability, logistics, behavioral economics.
- **Expression Style**: How you communicate and create. Writing, visual design, voice, prototyping, facilitation, teaching.

Once you have chosen, ask yourself: "What can I make or improve using this specific combination that someone with only one of these could not?"

Example:

- **Mindset**: Systems thinking
- **Tool**: Airtable
- **Lens**: Sustainability
- **Expression**: Facilitation

What this makes possible: Running workshops for procurement teams where you map their supplier network as a living system, track environmental data in real time, and help teams see where their decisions create downstream consequences they had not measured. No single specialist in that room can do all of that. You can.

Your stack will evolve. Revisit it every six months and notice what has changed: what you have deepened, what you have dropped, what new tool or lens has entered your orbit.

4. The Weekly Cross-Pollination Habit

What it is for: *Making polymath thinking a practice, not an event.*

The previous three frameworks require reflection and intention. This only requires thirty minutes a week and a willingness to endure a short period of unfamiliarity. Cross-pollination does not happen in annual strategy sessions. It happens in the small, repeated acts of encountering something outside your usual frame and letting it sit next to what you already know. Over time, those encounters become a habit of mind: you start noticing connections automatically, not just when you sit down to look for them.

Use this as a standing weekly commitment, not something you return to when you feel inspired.

Choose one small action each week:

- Watch a short video (under 20 minutes) from a field you do not work in. Not to become an expert, just to encounter one idea you would not have found otherwise.
- Ask someone from a different function what they are working on right now. Not small talk. Genuine curiosity about their problem.
- Open a tool you do not fully understand and spend twenty minutes exploring it without a goal. Confusion is productive here.
- Write or sketch one idea: a paragraph, a diagram, a rough outline that connects something from your past to something you are moving toward.

Thirty minutes. One action. No deliverable required.

What you are building is a reflex. The goal is that, eventually, you stop needing to schedule this because you do it constantly: in meetings, in conversations, in how you read the news. That is when you know the polymath orientation has become part of how you think, not just something you practice.

The difference between someone who reads about polymaths and someone who becomes one is almost always this: they kept showing up to small experiments until the experiments became a way of life.

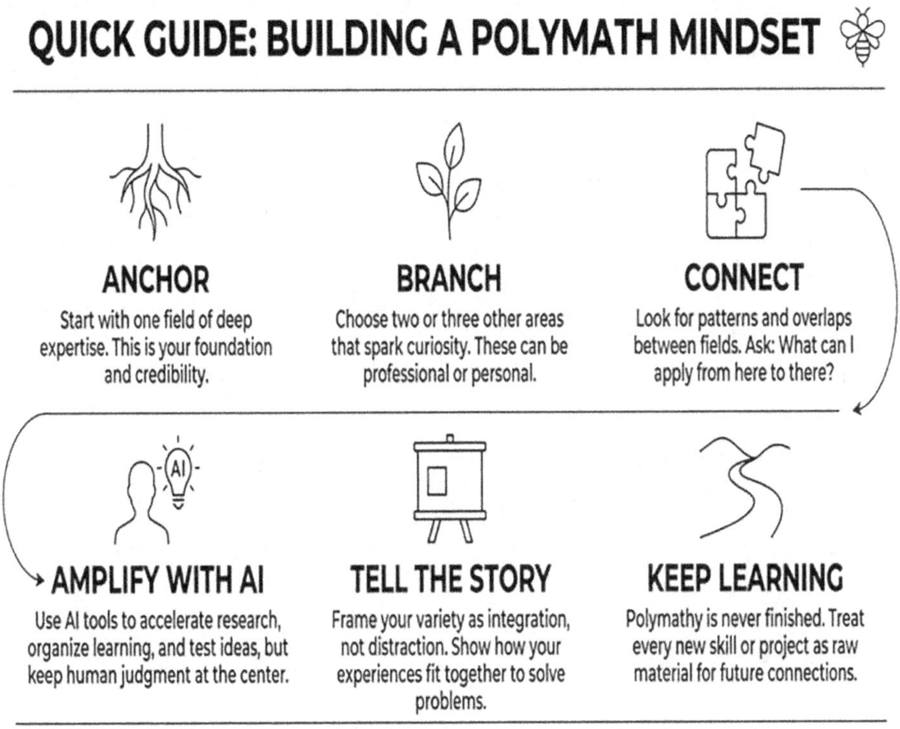

Figure 4 - Quick Guide to Building a Polymath Map.
Created with Nano Banana PRO (2026).

From Polymath to Method

The Polymath Approach expands what is possible. It builds range, pattern recognition, and the ability to connect ideas across domains. But range alone does not guarantee movement. Without a method, polymathy remains as potential, interesting, inspiring, and still scattered.

ADAPT is the structure that turns a portfolio of skills into a repeatable way of navigating change.

The five elements are not five separate concepts, and they are not a menu where you pick the ones that appeal to you. They work in sequence because each step sets up the next. Awareness is the radar that catches weak signals early and gives an honest read of what is changing. Direction turns that signal into a choice about where to aim. Action converts the choice into proof through small, low-stakes experiments. Positioning makes that progress visible to others, so it translates into trust, opportunity, and momentum. Tech Empowerment reinforces the whole system by reducing the effort involved and making the new way of working easier to repeat.

The chapters that follow slow each element down and show how it becomes usable in practice. The shared goal across all five is the same: moving from abstract understanding to practical execution, so that adaptation becomes a habit rather than a one-time scramble. Chapter 5 begins with Awareness, because noticing earlier than others is the simplest competitive advantage in these early stages of the AI era.

The Five Elements

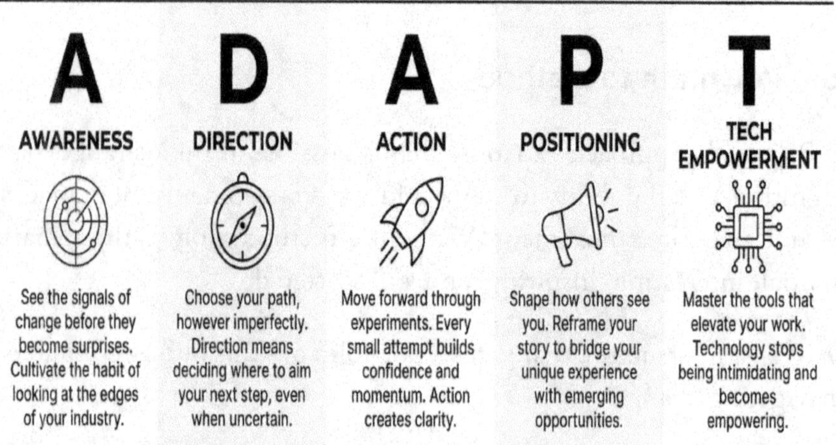

Figure 5 - The ADAPT Method. Made with Nano Banana PRO (2026).

In the last few years, researchers, educators, and regulators have started to agree on something important: people do not become "AIliterate" just by playing with tools in an unstructured way. Emerging AI literacy frameworks and formal AI literacy programs show that when learning is organized into clear stages and practices, individuals are far more likely to build the mindset, skills, and confidence needed to use AI responsibly and creatively in their work. Under the EU AI Act, for example, organizations are expected to design structured AI literacy programs rather than leaving employees to figure AI out on their own, precisely because a systematic approach is needed to manage risk and build capability at scale. ADAPT is my version of such a framework: a simple, methodical way to move from vague curiosity about AI to an intentional practice of noticing change, choosing a direction, acting, repositioning yourself, and, ultimately, feeling empowered by technology instead of threatened by it.

A: AWARENESS

Everything starts with Awareness. It is the act of paying attention, of seeing the subtle signals of change before they become a surprise. For me, that moment came late one night in 2022 when I first used ChatGPT to analyze a contract. It wasn't just a new tool; it was a signal that the nature of knowledge work itself was about to be rewritten. That single experience moved my awareness from passive curiosity about AI to an urgent need to understand and adapt. Awareness is the habit of looking beyond the center of your industry toward what is forming at its periphery, paying attention to emerging technology, and asking, 'What does this mean for me?' Without it, adaptation is impossible; with it, you can start to see the future taking shape in the present.

By simply tracking exchange rates daily and sharing them with managers, I solved a problem no one else had spotted. That awareness positioned me as more than a trainee; it made me useful. Years later, the same principle applied to AI. I wasn't the most technical, but I noticed where AI could save time and create insight, and that awareness became my edge.

Awareness in Action (DuPont, late 1990s): My formal role was to migrate purchase orders into SAP. But while doing that work, I noticed that every buyer on the team was manually pulling the same exchange rates from newspapers each morning, independently, to arrive at the same numbers. Nobody had flagged it as a problem because it had always been done that way. I started compiling the rates once and distributing them to the whole team. That small act of noticing, not a directive or a formal initiative, was my first real lesson in what awareness means: paying attention to what is already visible to everyone but seen by very few.

D: DIRECTION

Awareness can overwhelm and be overlooked. It is easy to see all the changes happening and freeze. What saved me in those moments was choosing direction, however imperfectly. When I lost my savings in building Kindmates, the dog food business, awareness alone was not enough. I knew the industry had shifted, that my timing was off, and that the pandemic had ended the opportunity. But direction meant deciding what to do next. The choice to knock on Unilever's door again, after eight years away and six selection processes, was not about returning to the familiar. It was about redefining my path in light of the coming AI wave. Direction is less about knowing every step ahead and more about committing to where the next one should go.

Direction in Action: After years in procurement, I could feel the fatigue setting in. The safe choice was to stay put. But direction meant enrolling at the Open University at 36 and later pursuing a master's degree at the University of Wisconsin at 43. Each decision gave me a new horizon, even when it felt uncertain. Those pivots prepared me to later choose AI, a path I hadn't imagined for myself before.

A: ACTION

The hardest part is moving. Experimenting with early AI tools came before any certainty about where it would lead. Action creates clarity. Every small experiment, every messy attempt, even if you do not have a clue what you are doing, builds confidence and momentum. This was how I shifted from being a procurement professional watching AI from the sidelines to someone leading projects and eventually joining the AI team itself. The Shark Tank win at Unilever is a perfect example. I did not wait for someone to hand me an AI project; I pitched one. I built a mockup, gathered a team, and showed what was possible. That action did not just win funding: it changed the way people saw me, and the way I saw myself.

Action applied (Laporte Chemicals): I could have kept issuing purchase orders manually, as everyone else did. Instead, I taught myself Microsoft Access and built a mini-ERP. That small action not only saved time but also showed my managers I could solve problems differently. Later in life, this same bias toward action helped me test AI tools without waiting for permission. Action has always been the lever that moves me forward.

P: POSITIONING

It is not enough to learn in silence. You must shape how others see you. My credibility did not come from a degree in computer science; it came from how I positioned my story. From LinkedIn posts to internal presentations to simply how I described my role in conversations, I had to reframe myself as someone who could bridge business and AI. The early days at DuPont gave me a lesson here. When I delivered the first spend analysis using endless piles of green-and-white printer paper, I wasn't just crunching numbers. I positioned myself as someone who turns chaos into clarity. Years later, positioning became even more important when colleagues and leaders had to trust me to lead in an area where I had no formal pedigree.

Positioning in Action (Unilever AI Champions): When I joined the AI Champions network at Unilever, I knew I wasn't the 'tech guy.' My background was procurement. But by openly sharing my learning journey through internal workshops, LinkedIn posts, and AI Moments, I positioned myself as a bridge between the technical and the practical. Positioning did not mean faking expertise.

It meant showing how my unique career story gave me credibility to speak to non-technical colleagues.

T: TECH EMPOWERMENT

Finally, there is the moment where technology stops being intimidating and becomes empowering. I did not need to master every algorithm. I needed to find the tools that worked in my world and use them with confidence. Tech empowerment is about control; it is choosing the apps, platforms, and systems that elevate your work, rather than letting them overwhelm you. For me, this began in the 1990s with WordStar and DOS at Securit, then SAP migrations at DuPont, then Microsoft Access at Laporte. Each time, I found myself learning just enough to make the system work for me. The same pattern repeated with AI. I did not need to build the models; I needed to learn how to prompt them, integrate them into procurement workflows, and turn them into allies rather than threats.

5

AWARENESS

Seeing What Others Might Miss

A respected senior buyer served as the packaging negotiator for a consumer goods company. He knew every plant manager and supplier by name. When an e-sourcing pilot was announced, he joked that real negotiations happen over coffee, not screens. He skipped the training. For a while, nothing changed. His calendar stayed full, his relationships stayed warm, and his results stayed within range.

Then, global procurement rolled out e-auctions and centralized category strategies. Suppliers moved their best pricing to the auction lanes. His phone calls lost leverage in a single planning cycle. He still had relationships, but the pricing decisions were now being made by teams who lived in the data. He went from dealmaker to contract administrator almost without noticing.

What he missed was not a lack of intelligence or effort. He missed that the scorecard was moving from relationship to transparency. He read the change as a tool trend, not a power shift. After e-sourcing scaled, he managed to keep his job for a while. His calendar still looked full, and routine did not change much on the surface. But the real sourcing choices were now made by a data-focused team that presented scenarios rather than anecdotes. He became the person people copied for awareness, but did not invite to set a strategy.

That story resonates with me because similar patterns have appeared across various industries and roles over many years. While the specifics may vary, the overall pattern remains the same. A professional develops skills within a specific way of working, and over time, that skill becomes second nature. This comfort turns into an unnoticed routine. Meanwhile, during this familiarity, changes in the environment occur that the routine was never meant to recognize.

The effects show up in practical ways before they show up on paper. A person can lose influence without losing a title. Strong effort produces weaker results. A role evolves while the person in it is still working from an earlier version of the job. The signs are usually easy to spot once you know what to look for. Someone keeps improving an old process without asking whether it still matters. A professional network is treated as permanent rather than something that needs to be renewed. Reports focus only on what has already happened, with little attention to what may be coming next. New tools are ignored until the company creates a formal program around them. An approach that delivered results in the past is defended even as the environment around it keeps changing.

None of this makes a person incompetent. It makes them unaware. And in a professional world being reshaped by artificial intelligence, the distance between unaware and unprepared is shorter than it has ever been.

Awareness starts with recognizing that pattern, not in the abstract, but in the specific context of your own career. Many professionals do not become less relevant because they lack talent. They drift because routines that once worked become hard to question. Familiar ways of working feel efficient, dependable, and safe, so people keep following them even after the context has changed. In the age of AI, that pattern is not harmless. Adapting is no longer a choice reserved for the ambitious. It is a requirement.

ADAPT is a framework for responding to change with intention. It begins with Awareness because, without awareness, the rest is guesswork. You cannot choose a direction if you do not understand what is changing. You cannot act well if you do not know which problems matter most. You cannot reposition yourself if you do not see what the people and organizations around you now value. In the AI era, awareness is the starting point for everything else in this framework.

However, awareness isn't usually a sudden discovery. It more often resembles how your eyes slowly adapt in a dark room. The shapes have always been present; you simply weren't paying close enough attention.

The Instinct to Act Before Having All the Answers

A talented category manager had everything under control, or so it seemed. The high-volume packaging resin she managed came from a primary supplier that had never given cause for concern. Quality was excellent, costs were low, and shipping routes were predictable. When news alerts about a citywide lockdown in Wuhan, China, the first major signal of the coming COVID-19 pandemic, started appearing on her screen, she dismissed them as operational noise.

Port congestion alerts followed, but she decided to postpone qualifying a second supplier until after the quarter closed. Her goal was to protect her department's savings targets, and on paper, the delay seemed reasonable.

Within six weeks, what had started as a distant news story became a supply crisis with direct consequences for her portfolio. The shipping lane from her supplier shut down. Lead times for new orders doubled, and manufacturing plants that depended on the resin had to start rationing stock. To prevent a production halt, the finance department authorized emergency air freight, a solution exponentially more expensive than sea transport. The increase in shipping costs eliminated two years of negotiated savings. Her reputation for reliability took a serious hit, and despite her efforts to manage the situation, the label that followed her was slow to escalate.

What she missed were the weak signals that her seemingly stable supply base was vulnerable. Running scenarios early enough to still have options is what awareness produces. She had managed her part of the system well but had not stepped back to assess the system as a whole, and by the time the signals were impossible to ignore, the options had already run out.

This is how awareness tends to work in practice. The signals are not hidden. They appear in trade publications, in news feeds, in the questions that clients and colleagues start asking, in the subtle shifts in what gets funded and what gets cut. The difficulty is not access. It is attention. It is the willingness to look at something before being forced to.

The packaging buyer who lost leverage to e-auctions missed an industry signal, a broad structural change that was redefining how an entire sector operated. The category manager missed a different kind of signal, one that was found where geopolitics and supply chain fragility converge. Both signals were visible. Both were ignored because the current way of working still felt adequate.

A third type of signal is even subtler and tends to surface within your own organization before anywhere else.

The Team That Fell Behind by Being Careful

Through a colleague I interviewed for this book, I heard about a pattern that had unfolded inside the large pharmaceutical company he works for. A team lead there had built a solid reputation over several years managing a strong group of analysts. When Microsoft Copilot and ChatGPT became available across the organization, she responded that many leaders in regulated environments do so with extra caution. Questions about data handling, compliance boundaries, and appropriate use cases remained unsettled, and waiting for clearer guidance seemed prudent.

A peer team in a similar function took a more experimental approach. Operating under the same policies, they began testing AI on lower-risk internal work, such as RFP (Request for Proposal) outlines, contract markups, and market summaries. The early results were not perfect, but somehow useful. Over the next six months, their turnaround times improved, and stakeholders increasingly began sending work their way because responses came faster and draft outputs were ready sooner.

When the function was later reorganized, that peer lead moved into a newly created AI leadership role. The other team continued to play an important role, but its work increasingly focused on approvals, reviews, and escalations rather than on shaping new workflows.

No one was singled out, and no one was judged to have failed. The change was subtler than that. Stakeholders had already started to change their habits based on where they were getting speed and responsiveness. By the time leadership formalized the new structure, it was largely recognizing a change that had already taken place. The more cautious team had not done anything wrong, but it had become less central to the evolving work. It was less visible than the team that took a more pragmatic approach.

This is a different kind of signal. It is not about global markets or supply chains. It is about how the definition of a role changes while the title stays the same. A manager who once spent most of her time leading people now spends more time interpreting dashboards. A marketing professional who used to brief designers starts writing prompts for image generation. A procurement analyst who was valued as a domain expert started being evaluated as a data translator. The clue is where the cognitive load moves. When the repetitive parts of a job start getting automated, value migrates toward context, creativity, and connection, and the people who notice that migration early are the ones who get to shape it rather than react to it.

And then there is a third kind of signal, the most personal and the most ignored. It shows up when you encounter vocabulary or tools you do not yet understand. When colleagues began discussing "prompt chaining" or "fine-tuning models," I did not dismiss it as technical jargon. I wrote it down, looked it up, and tested a small version myself. That habit of tracing confusion to curiosity is how awareness becomes learning. Every time you notice someone working differently, faster, or in a way you do not fully understand, that is a signal. You do not have to master everything. But you do need to understand enough to collaborate intelligently.

Across these three stories, a single pattern connects them: mistaking activity for relevance. Heavy calendars hide weak signals. New tools are seen as optional efficiency gains rather than fundamental changes in how power and influence move. Small experiments are delayed until a formal policy is in place, long after the window of opportunity has closed. Resilience is treated as a nice-to-have rather than a core metric. Energy gets spent protecting yesterday's advantage instead of building tomorrows.

Lessons From My Own Missed Signals

This pattern is something I had to learn to recognize in my own career. If I had to pinpoint when I first developed an awareness mindset, it was not a single breakthrough moment. It was something that built up slowly, shaped by small realizations that the world of work changes long before anyone announces it.

I have had plenty of missed signals myself, and talking about them is more useful than pretending they do not exist.

Around 2008, while leading IT Procurement at Unilever Brazil, large IT vendors, especially IBM, were pitching Business Process Outsourcing to run procurement operations end-to-end for certain categories of materials and spend. I agreed with leadership that saw it as a cost play rather than a forecast. Two years later, shared services absorbed a big chunk of transactional buying across regions. None of this ended my career, but it was a wake-up call. When a supplier or other partner proposes running your function, assume they are reading a future you and your company will eventually move toward. My learning curve could have been avoided if I had treated that pitch as a strategic signal rather than a sales pitch.

While I was polishing how we did the work, the firm was deciding where the work should live. The better move was to put myself at the table when the target operating model was defined. That is what awareness does at its most useful. It moves you from the work to the design of the work.

While studying for a bachelor's degree in environmental studies at The Open University, I was researching sustainability and ESG (Environmental, Social, and Governance) reporting at a time when many companies still treated it more as a communications issue than an operating requirement. I saw it that way too. Later, as customer expectations shifted toward supplier traceability, human rights attestations, and audit-ready data, it became clear that this was no longer peripheral. It was becoming part of how companies were evaluated and how credibility was established. That is what a professional blind spot can look like. You may have the right experience and credentials, yet still be unprepared for the questions that matter most when expectations change.

Looking back, the problem was not a lack of effort or competence, but a failure to recognize that the questions shaping the future had already changed while I was still answering the old ones well.

How Awareness Developed Over a Career

The DuPont SAP migration I described in Chapter 1 was an early training ground for this kind of attention. At the time, most people around me saw the new system as a technical update. What I noticed was different: it changed who controlled the information. Suddenly, it was not the senior buyers with decades of supplier relationships who held the advantage. A few people understood how to navigate the new system. I realized that change does not arrive with a loud announcement. It often comes subtly, disguised as an IT upgrade. The daily exchange rate reports I compiled at DuPont, which I described in Chapter 1, taught me something similar. The task itself was tedious. But it gave me a front-row view of how currency moves could impact prices and margins. Without knowing it, I was learning to connect dots between small signals and big outcomes, a mindset that would later define how I approached every major change in my career.

When I arrived at Unilever in 2005, I inherited a department without a manager from my first day, a situation that forced me to operate at a level of alertness far beyond the formal role. I had to quickly make sense of unfamiliar situations, anticipate problems before they became more complex, and make decisions with very little guidance. That experience taught me that awareness is not only about spotting external trends. It is also about understanding the system you are working within and recognizing how much others rely on the quality of your judgment.

As the company globalized and centralized processes, I noticed another subtle pivot. The value of procurement was moving from pure price negotiation to insights, becoming more strategic in that sense. Those who could interpret data, not just quote prices, started influencing the conversation. I began experimenting with basic automation, such as Excel macros and spend analysis templates, and that small curiosity helped me stay relevant as digitalization began accelerating around 2010.

Then came the first sign that the traditional procurement rhythm was aging. Conversations around sustainability, traceability, and the circular economy started appearing in leadership briefings. Many peers saw them as PR trends. I saw the early outline of what would become a new business model, one that would later dominate discussions about supply chains.

By 2012, though, I was restless. The work had become predictable, and corporate stability felt like stagnation. I could sense that my trajectory no longer aligned with where I believed the market was heading, even if I could not yet articulate an alternative. Instead of waiting for clarity, I chose to experiment with entrepreneurship.

The Missing Years and What They Taught About Attention

Running my own ventures forced me to develop a different kind of awareness. There were no quarterly reviews or dashboards to tell me what was changing. I had to feel it. When you are an entrepreneur, awareness is survival. You see patterns in customer behavior, supply chain fragility, and the fine line between optimism and denial. Every conversation with a potential client became a signal about what the market was willing to pay for and what it was not. Every failed pitch taught me something about timing and readiness that no internal strategy document had ever made visible.

Those years taught me that awareness does not just protect you. It also, and most importantly, expands you. It makes you sensitive to context and timing, two qualities that would later help me navigate the AI wave. There is a particular kind of attention that entrepreneurs develop, a willingness to listen for what people need rather than what they say they want, and that skill translates directly into the corporate environment when you bring it back. The vocabulary is different. The underlying discipline is the same.

What also changed during those years was my relationship with uncertainty. In a large organization, uncertainty is something most people try to remove from their plans. As an entrepreneur, it is the medium you work in every day. That forced a habit I still carry, treating ambiguity as information rather than as a problem. When something feels uncertain, the question is not how to eliminate the uncertainty, but what the uncertainty is telling you about the situation.

When I returned to Unilever in 2021 after almost eight years away, the company felt different. The technology available was more advanced, but what caught my attention was not the tools. It was how much faster the organization was moving. The decision cycles were shorter, the language was more data-driven, and expectations for productivity had multiplied. I could sense that another transformation was brewing, even before AI became the headline. Coming back after years outside gave me a strange advantage: I was not numbed by the gradual pace of internal change. I was seeing it fresh, and the gap between where the organization was and where the technology already pointed was obvious to me in a way that it might not have been if I had never left.

Long before ChatGPT entered mainstream conversation, there were signals inside Unilever that something larger was forming. By 2021, procurement teams had automated many repetitive processes. Bots processed purchase orders and reconciled invoices with impressive efficiency. Yet the core of the work, judgment around spend allocation, supplier risk, and forecasting, remained largely manual. We had automation, but not intelligence. I began to question how long it would take before predictive capabilities, already visible in marketing and finance, would move into procurement. The gap felt temporary.

At the same time, analytics was expanding rapidly. Dashboards multiplied. Reports became more sophisticated. Data was no longer scarce. Interpretation was. Teams were producing more metrics than ever, but translating those metrics into decisions still required slow human synthesis. The imbalance was obvious: information was accelerating, but insight was not. That tension almost never resolves on its own, and it usually invites a technological solution.

Another signal came from observing how people were learning. Some colleagues were not waiting for formal training programs. They were experimenting with digital tools, enrolling in short online courses, and testing ideas in real time. Years earlier, I had learned SAP through hands-on trial rather than structured instruction. The instinct to experiment was resurfacing, only now the tools were more powerful and more accessible.

Perhaps the strongest signal, however, was cultural rather than technical. When artificial intelligence was mentioned in leadership meetings, it was often dismissed as something relevant "in the future." I had seen that posture before during early ERP implementations and later during sustainability debates. Dismissal is almost never about irrelevance but about sensing implications that feel easier to delay than to face. The language people use when they dismiss something tells you more than the language they use when they embrace it. "Not yet," "let's see," and "it's too early" are rarely assessments. They are forms of self-protection. The underlying message is not that the technology is unready, but that the person is not ready for what it implies.

By the time ChatGPT became widely available in late 2022, the groundwork for that moment had already been laid. The automation of repetitive tasks, the proliferation of dashboards, the growing gap between data and insight, the experimentation happening at the edges, all of it pointed in the same direction. What surprised many was not the arrival of AI, but the speed at which it became usable. For those paying attention, the pattern had been visible for at least two years. The tools were catching up to a need that had been growing in plain sight.

Awareness, I realized, is not about predicting specific technologies or claiming foresight. It is about noticing resistance early. It is about recognizing when processes feel slower relative to available tools, when language changes inside meetings, and when phrases like "we have always done it this way" begin to sound defensive rather than confident. Those moments are almost never obvious in isolation, but they accumulate into a pattern that eventually becomes impossible to ignore.

What Gets in the Way

If awareness is mostly about paying attention, why do capable people keep missing the signals? Because the obstacles are not intellectual. They are emotional.

The most common one has nothing to do with ability or access. It is time, or more precisely, the feeling that there is none. Meetings, emails, and deadlines leave no space to look beyond what is due today. Ironically, those who are busiest are often the most at risk of missing the changes that will redefine their jobs. Being busy creates a convincing sense of forward motion, but it also crowds out the kind of attention required to notice new things.

I lived this during my years in Switzerland, the same period I described in Chapter 2. My calendar was full, my results solid, and in theory, everything looked fine. But I was not learning. I had mistaken constant activity for growth. When I finally stepped back, I realized my work had not changed in years. Only the volume had. The way I eventually broke that pattern was by treating thinking time with the same seriousness as a meeting. A short block each week for scanning and reflection. Once it was on the calendar, it stopped feeling like an indulgence and became part of the job.

Another obstacle is the pride that comes from knowing how things work. Expertise is valuable, but when it goes unquestioned, it can restrict how you see new problems and new possibilities. The more competent you are, the easier it is to believe that your way is still the best way. You stop asking naive questions because you do not want to look uninformed.

I remember this clearly from my early years in procurement. I had mastered negotiation playbooks, supplier scorecards, and every KPI (Key-Performance Indicator) that mattered. Then automation started creeping in. My first reaction was not curiosity. It was skepticism. I told myself, "These systems will never replace judgment." I wasn't wrong; I was defensive. That is how resistance starts: not in ignorance, but in confidence. What shifted for me was reframing learning as part of my value rather than a threat to it. Instead of proving I already knew, I started proving I could adapt. The minute I saw learning as reputation-building rather than reputation-threatening, curiosity came back, and things started to flow more easily.

Denial shows up, too, though it is not always obvious. Sometimes it appears as jokes or quiet disinterest. I have seen very smart professionals brush off technology by saying things like, "Let's see if it lasts," or, "It is just a trend." I have said those things myself. It is easier to look away than to decide what to do next. It allows you to keep your identity intact a little longer. The cost of that comfort is invisible at first, because you are still performing well by the standards that existed before the shift. The cost becomes visible only when the standards themselves have moved, and you have not.

When I first heard colleagues talk about AI-powered sourcing, I dismissed it as another consulting buzzword. At a steering committee meeting to review resiliency, a vendor presented a tool that claimed it could scan thousands of suppliers and surface risks in minutes. I nodded politely and moved on. I had spent years building intuition, relationships, and judgment. I do not believe a model can replicate that. Part of me also did not want to believe it, because if it could, what did that say about the two decades I had invested in learning to do it myself?

A few weeks later, one of my team members ran the same tool on a limited category review. The output was not perfect, but it was fast. It flagged supplier concentration risks I had not prioritized and surfaced alternatives I would not have considered until much later. The numbers were hard to ignore. Cycle time dropped. Accuracy improved. What I only understood later was not what the tool could do, but how quickly I had resisted it. That reaction had very little to do with evidence and a lot to do with what it implied for me personally, because I was worried about becoming irrelevant in a field I had spent two decades learning. Once I saw that, it was clear the unease was less about the technology and more about identity.

The way out was simple and repeatable: test the thing I wanted to dismiss. When resistance showed up, I ran a small experiment instead of arguing with myself. I used the tool for a narrow task, read the underlying paper rather than reacting to the headline, and asked questions before defending a position. Most of the time, that process produced something useful. When it did not, it still clarified what was not ready yet, and that clarity mattered. Over time, testing replaced fear with data, making the whole conversation feel more grounded.

And then there is the problem of volume. The modern professional is overwhelmed by information. Articles, podcasts, newsletters, webinars. The confusion comes from treating consumption as awareness. More input does not produce better judgment. It often produces hesitation.

When everything looks relevant, action slows. Research in *Nature* reveals a fundamental "additive bias," where the human brain systematically overlooks the option to subtract, defaulting instead to adding more layers. True clarity demands we fight this instinct: fewer sources, clearer signals. A small number of trusted inputs matters more than constant scanning. The discipline is not reading more but deciding what to ignore. Awareness is built through synthesis, not volume. It comes from connecting what matters to action.

Finally, awareness often clashes with ego. After building a career on competence, admitting your skills need updating can feel like a personal critique. I experienced this after my entrepreneurial failure. For months, I blamed timing and market conditions, only later realizing I had ignored signs that the infrastructure and customers were not ready. I wanted the idea to work, so I filtered out contradictory data. What changed was separating failure from shame. Once I treated mistakes as data rather than identity, I could see them clearly. Awareness thrives when you stop needing to be right.

A FOR AWARENESS: THE COST OF MISSING SIGNALS

Career comfort creates missed signals. Small changes ignored today become crises tomorrow.

THE 3 SIGNALS TO WATCH

- **Industry & market:** New business models, regulation shifts, fast-rising entrants

- **Role evolution:** Where the cognitive load is moving in your job

- **Skills & capability:** New terms, tools, and methods you do not yet speak

WEEKLY AWARENESS PRACTICE (30 MINUTES)

- **Scan 10 min:** Use one research or AI tool to surface fresh changes. Note repeats and surprises

- **Reflect 10 min:** What gets easier or irrelevant if this grows? Where do I see the same pattern at work?

- **Act 10 min:** Do one move now. Share an insight. Test a feature. Bookmark a deep dive. Start a tiny experiment

ADAPT Framework • A is for Awareness • Part 1 of 3

Figure 6 - Awareness. Made with Nano Banana PRO (2026).

The obstacles to awareness are rarely about information. They are about identity, comfort, and the emotional cost of looking honestly at what is changing. Recognizing these patterns in yourself is not a weakness. It is the beginning of a different kind of attention.

Adapter Profile: Clara, The Teacher Who Became a Learning Designer

Clara taught high school history in Ohio for sixteen years. Her Advanced Placement (AP) pass rates, referring to college-level exams taken by high school students, were well above the national average. Then, during exam season, she noticed a change: students were turning in essays that sounded different. The grammar was better, the arguments more organized, but the writing no longer sounded like their own.

She could have banned AI tools, but she got curious instead. She spent several days using ChatGPT on her own essay prompts. She graded the outputs using her rubric. Some earned a B+. None earned an A. And none demonstrated genuine thinking.

Clara realized that if AI could produce acceptable work, her role was no longer to teach competence. It was to teach the kind of thinking AI could not easily reproduce, such as personal interpretation, source criticism, and connecting history to lived experience. She redesigned her curriculum around "primary source investigations," in which students interviewed family and community members and placed their stories in historical context. AI could imitate that work if students created false sources, but it could not genuinely reflect the insights that come from real human conversations and a firsthand perspective.

What made Clara's response unusual was not just what she did but how quickly she did it. She did not wait for the school board to issue a policy. She did not attend a conference on AI in education before forming an opinion.

She ran her own experiment, using the same tool her students were using, and she let the results tell her what needed to change. That speed came from awareness. She noticed a signal, a shift in the texture of student writing, and instead of reacting defensively, she followed the signal to see where it led.

By the next school season, Clara was leading a district-wide working group on AI-responsive pedagogy. That summer, an education technology company recruited her to help design AI-augmented learning tools. Today, she works as a learning experience designer, reaching more students than she ever could from a single classroom. Her salary increased by more than forty percent. Her working conditions improved. And the expertise she had built over sixteen years of teaching, the deep understanding of how students learn, became more valuable in the new context than it had been in the old one.

When I asked what made her move, she said, "I noticed I was spending more energy policing the tool than understanding it. That is when I knew something had to change."

That sentence is what awareness looks like in practice. Not a grand strategic pivot. Not a multi-step diagnostic process. A single honest observation, followed by the willingness to act on it. Clara lacked a framework for awareness. She had a habit of paying attention to what was happening in front of her, and when it did not match her expectations, she adjusted them rather than defending them.

What a Weekly Practice Actually Looks Like

At some point in writing this chapter, I realized I should describe what awareness actually looks like in my life rather than present it as a pristine system. The version I practice is simple, and it has changed shape several times over the years. Most weeks, I block a short window, usually Friday morning, sometimes Sunday evening, where I step back from execution and look at what has changed. My version of this has three loose parts, though I would hesitate to call them steps, since they often blur into one another.

The first part is scanning. I spend about ten minutes reading across a few sources that I trust. Tools like Perplexity, Feedly, and LinkedIn Pulse make it easier to spot trends across multiple industries without requiring me to visit 20 different websites. As of early 2026, Perplexity is the easiest to use because topics can be prioritized directly on the home screen. Research summaries from sources like *MIT Technology Review*, *Harvard Business Review*, *McKinsey*, and *CB Insights* can add depth when I want to go further. But the main value is not the news itself. The value lies in pattern recognition: repeated themes across different outlets, such as automation in customer service or the rise of multimodal AI models, that signal momentum rather than noise. I am not reading for depth at this stage. I am reading for repetition. When the same theme shows up across different outlets, different industries, and different conversations, that convergence means something.

The second part is connecting. Another ten minutes, usually with a coffee, asking myself a version of the same question: does any of this connect to something I am already seeing at work? If a trend grows, what part of my job becomes easier, and what part becomes less relevant? I write short answers. The writing is what turns information into awareness. Without it, the scanning stays passive. Sometimes the connections are obvious. Other times, I notice something that does not quite fit but stays in my mind, and that lingering sense of something unresolved is often more valuable than the connections that arrive easily.

The third part is acting. I'll do one small thing based on what I noticed. Share an insight with a colleague and ask their reaction. Test a new AI feature I read about. Bookmark something that answers one of the questions I have been carrying. The action does not have to be impressive. It just must exist, because awareness without any movement fades into trivia. Some of the most useful actions I have taken were conversations. A five-minute exchange with a colleague in a different department, asking what is frustrating them lately, has sometimes told me more about where the organization is heading than any report.

In total, this takes about thirty minutes a week. Over a year, that adds up to roughly a full workweek of strategic reflection, something most professionals never do. I do not always hit the full thirty minutes. Some weeks it is ten. Some weeks, I skip it entirely because work is too demanding. What matters is that the rhythm exists, and when I return to it, the habit is still there.

A FOR AWARENESS: THE SCANNING LOOP

Awareness is a discipline, not luck. Build it through regular practice.

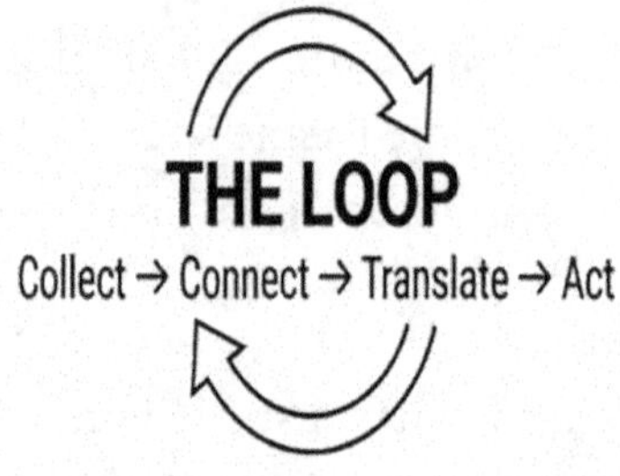

Figure 7 - Awareness Scanning Loop. Made with Nano Banana PRO (2026).

The scanning loop is a rhythm: read for patterns, connect them to your work, and act on one thing before the weekend. The loop compounds over time, not because any single week produces a revelation, but because consistent attention builds a kind of peripheral vision that most professionals never develop.

If you are already doing some version of this, even informally, even without calling it anything, that counts. You do not need a formal system. You need consistent attention. The signals are already arriving. The question is whether you are giving yourself enough space to notice them.

The good news is that much of this can now be supported by simple agentic AI tools that I will cover in later chapters. So do not be put off by the amount of effort this may seem to require at first. Much of it can be set up once and then managed with an agent's help. That is one of the main ideas behind this book. For anyone who wants a more structured version, including a three-column exercise for mapping blind spots, guided prompts for each signal type, and a quarterly awareness audit, I maintain those at adapterslab.ai/awareness. But the structured version is optional. The habit is what matters.

The Awareness Advantage

We live surrounded by insights, reports, and data dashboards that promise clarity, yet awareness alone rarely translates into action. People often sense something changing long before they can explain it, but once they see it clearly, they also see the responsibility that comes with it. Acting early means moving before others do, without certainty or approval, and that is almost never easy.

Waiting can feel like the responsible choice, even a sign of good judgment. But there is a difference between holding back for a real reason and simply hoping that something will resolve itself, that the organization will catch up, or that the right moment will hold long enough. One is a decision. The other is avoidance, and it almost never leads anywhere new.

The good news is that awareness does not need to become heavy or overwhelming. Much of what I have described in this chapter, the scanning, the connecting, the small weekly habits, can now be supported by simple AI tools that I will cover in later chapters. You do not need to build an elaborate system. You do not need to become a futurist or a trend analyst. You need a steady rhythm of attention and the willingness to look at what you find, even when it challenges your assumptions about how things are going.

Throughout my career, every significant change, from procurement to entrepreneurship to AI, started with a feeling of discomfort. It wasn't a crisis, but a persistent sense that the work was becoming too routine, that the focus of leadership questions was shifting while my responses remained the same, and that some colleagues were heading in unfamiliar directions. Although this tension was easy to overlook, it often held essential insights. Once I learned to interpret it rather than resist it, it outlined what needed adjustment. Now that AI can process and organize information faster than any person, the advantage lies not in knowing more but in recognizing sooner. Awareness is the capacity to pay deliberate attention, to see patterns before they turn into real problems, to notice when something no longer fits, and to recognize openings while they can still be reached.

This is why Awareness forms the foundation of the ADAPT method. Without it, direction becomes uncertain and action reactive. With it, adaptation becomes intentional. And the distance between those two states, between reacting to change and anticipating it, is often no more than a few months of consistent attention.

Awareness does not prevent disruption. It does not guarantee an advantage. But it shortens reaction time. For me, it began with curiosity about how systems operate, evolved into observing how influence flows through them, and matured into recognizing how technology redefines roles before job descriptions are updated. It is disciplined attention, not intuition.

Before You Take the Next Step

Reinvention starts earlier than most people expect, with awareness and a transition in how decisions are made. Looking back across the career arc described in Part I, every turning point began with awareness, even when I did not realize it at the time. The DOS machine at Securit, the SAP migration at DuPont, the move from traditional negotiation to data-based decisions later in my career, the AI shock in 2022, each of those moments came from noticing small signals before they grew into surprises. That is the real power of awareness. It does not make you immune to disruption. What it does is give you enough time to respond deliberately rather than reactively.

If you remember nothing else from this chapter, let it be the five core principles that form the foundation of true awareness. They are not complex, but they require consistent attention.

BUILDING AWARENESS

A Skill You Build	Signals Show Up	Busyness Hides	Curiosity Helps	Leads to Choice
Awareness is a skill you build. It comes from small, regular habits, not sudden bursts of motivation.	Useful signals show up all the time. They are easy to miss unless there is space to slow down and pay attention.	Being busy can hide what matters. When everything is urgent, it is easy to confuse activity with real progress.	Curiosity helps when things feel threatening. When the instinct is to judge or defend, try asking one honest question first.	Awareness has to lead to a choice. Noticing change is only helpful if it informs what to to do next.

Figure 8 - Building Awareness. Made with Nano Banana PRO (2026).

These five principles are the "what" and the "why" of awareness. They are not tasks to complete. They are habits of attention that develop through repetition, the kind of repetition this entire chapter has been describing through stories, mistakes, and small adjustments made over the course of a career.

This is true whether the signal is an industry-level shift, like the move from relationship-based procurement to data-driven platforms. Or a role-level change, like the pharmaceutical team lead whose caution cost her visibility. Or a personal skill signal, like encountering vocabulary you do not understand and choosing to trace the confusion rather than dismiss it. All three types follow the same logic: notice early, interpret honestly, and respond before the window narrows.

You are probably already doing more of this than you think. The fact that you are reading this book is itself a form of awareness. The next step is not a dramatic overhaul. It is giving yourself a small, regular window to notice what you are already sensing. No more information. Better attention to the information you already have.

Begin modestly. Set aside time this week to step back from your daily cycle of execution and look at what has shifted, what feels out of place, and what you have been postponing. The signals are already there. The question is whether you are ready to notice them.

Awareness is the lens that brings the changing world into focus. It allows you to see the waves of disruption before they arrive. But seeing the wave is not enough.

In the next chapter, we will move from observation to decision. We will explore Direction: how to choose your path when the future feels uncertain and how to build a career that aligns with both your strengths and the world's new reality. Because awareness tells you that change is coming. Direction decides what you will do about it.

A FOR AWARENESS: FROM INSIGHT TO ACTION 

Remove blockers. Ask better questions. Turn signals into decisions.

⚡ COMMON BLOCKERS → QUICK FIXES

- **Busyness** → schedule a 30-minute signal block
- **Comfort of mastery** → ask one naive question per meeting
- **Denial** → test the thing you doubt
- **Overload** → keep 3 trusted sources, summarize the rest
- **Ego** → treat mistakes as data, not identity

🧭 STARTER QUESTIONS (YOUR COMPASS)

- **Industry:**
  Which technologies are reshaping how we create value?
- **Role:**
  What in my job could be automated in 2 years?
- **Skills:**
  What are two capabilities peers are learning faster than me?

ADAPT Framework • A is for Awareness • Part 3 of 3

Figure 9 - Awareness: From Insight to Action.
Made with Nano Banana PRO (2026).

6

DIRECTION

Choosing Your Path When the Future Is Unclear

The signs of change are now clearly visible, but the harder question is what steps to take next. Many people tend to stagnate at this point, often because they embrace uncertainty when choosing a direction. Remaining in the same place may feel safer, even if it isn't, while moving forward requires making judgments without any guarantees. A former colleague was offered a role abroad with significant visibility and responsibility. It was the kind of opportunity that could redefine a career and life trajectory. Instead of rushing to a decision, he carefully compared cost structures, schools, taxes, and long-term scenarios, perhaps with too much rigor. Weeks passed as he continued pursuing the perfect scenario with that structured, or should I say, over-the-board evaluation. By the time he was ready to commit, the role had been filled. The analysis was sound; however, the timing did not align, and he missed the opportunity.

That example highlights a pattern I've noticed repeatedly: paralysis disguised as diligence. Many professionals respond to uncertainty by gathering more information. They read, compare, consult, and model scenarios, waiting for a moment when the right path becomes obvious. That moment rarely happens. Instead, the list of considerations grows longer, options multiply, and advice from different directions begins to conflict. What started as careful thinking ends up being a reason not to decide. The analysis was never the issue. The real problem was using it as a substitute for making a choice.

Research supports this dynamic. In *The Paradox of Choice*, psychologist Barry Schwartz shows how an abundance of options can produce anxiety and delay rather than better outcomes. The effect appears in everyday situations. Standing in a supermarket aisle, confronted with multiple nearly identical products and feeling pressure to choose the right one. The stakes are trivial, yet the hesitation feels real. When the stakes increase, the effect intensifies. Career transitions. Relocation decisions. Investment in new skills. Information accumulates, but clarity does not.

Artificial intelligence intensifies this tension by accelerating the evolution of opportunities and roles, thereby shortening planning cycles. Skills that are important today might be automated sooner than expected, making waiting for complete certainty before acting seem responsible. However, in fast-changing environments, such hesitation often leads to stagnation as conditions continue to evolve. The graduate school decisions I described in Chapter 2, first the Open University, later the Wisconsin master's, both involved a choice between stretch and comfort. In each case, I chose the path that aligned more closely with prior experience. The sustainability route strengthened systems thinking and strategic awareness. What it did not provide was technical fluency. As artificial intelligence adoption accelerated, those technical foundations became increasingly valuable. The original choices were reasonable. Their trade-offs only became visible later. The stretch that had been postponed eventually returned.

A third form of hesitation emerges in the belief that clarity must precede movement. The entrepreneurial years and the return to Unilever, as described in Chapters 2 and 3, illustrate this clearly. Clarity did not arrive in advance. Direction became clearer only after engaging directly with a concrete supply crisis and working through the operational realities on the ground. Action generated information that reflection alone had not produced.

This pattern has a name in the research literature. Saras Sarasvathy, a professor at the University of Virginia's Darden School of Business, spent years studying how expert entrepreneurs make decisions under conditions of genuine uncertainty, not the manageable kind where probabilities can be estimated, but the deeper kind where the relevant markets, firms, or roles do not yet exist and therefore cannot be predicted at all. Her research, published in the Academy of Management Review, drew on detailed cognitive studies of founders who had built companies from nothing and introduced a distinction that challenges much of what professionals are taught about planning: the difference between causation and effectuation.

Causation is the logic most of us were trained in. You define a goal, analyze the environment, select the optimal strategy, and execute. It works well when the future is reasonably predictable, when you are choosing between known options with estimable outcomes. But Sarasvathy found that expert entrepreneurs almost never operate this way in the early stages of building something new. Instead, they use what she calls effectuation: they begin with available means, who they are, what they know, and whom they know, and move forward by taking small, concrete actions without a fixed end goal. They do not try to predict the future. They focus on controlling what they can. They commit only what they can afford to lose, a principle Sarasvathy calls "affordable loss."

They treat surprises not as deviations from a plan but as new data that reshapes the path. And rather than selecting the "optimal" strategy from a set of predetermined alternatives, they build partnerships and seize contingencies as they arise, allowing the venture, or in our case, the career direction, to take shape through the process itself.

As Sarasvathy writes, "the process of effectuation allows the entrepreneur to create one or more possible effects irrespective of the generalized end goal with which she started." The destination is not fixed in advance. It emerges from the interaction between what the person brings and what the environment offers. This is not a theory about winging it. It is a theory about a different kind of rigor, one that prioritizes action-generated learning over prediction-based planning.

The relevance to anyone working through a career shift during the current AI disruption is direct. In stable, well-mapped environments, prediction works. You can study the job market, identify growing fields, invest in the right credentials, and execute a five-year plan. But the environment most professionals now face does not operate that way. Artificial intelligence is rewriting roles faster than any strategic plan can anticipate, with new tools emerging daily.

Entire categories of work shift in relevance within a single quarter, and in this kind of environment, the effectual mindset is not just useful but may be necessary. You do not need to see the entire path, just take a step with the means you already have, observe what happens, and let the next step reveal itself through what you learn.

The colleague who over-analyzed the role abroad applied causal logic in a situation that required effectual logic, viewing the decision as a prediction problem when what was really needed was a willingness to move with imperfect information and adjust from there. Across these examples, a consistent pattern emerges as direction rarely arrives as a fully formed answer but instead develops gradually through movement.

Two Principles That Make It Manageable

Two principles helped me find direction when certainty was not available.

The first is to favor small, reversible moves over large, irreversible ones. Instead of committing to a hard change right away, try a smaller version first: a focused project, a short course, or a prototype applied to a real problem.

When interest in artificial intelligence intensified, the transition did not begin with a career change or enrolling in another degree program. It began with structured experimentation and practical application. Hosting short internal "AI Moments" sessions required testing tools, building small prototypes, and translating emerging technology into operational value for colleagues. The initiative was modest in scale, but it created accountability, feedback, and visible proof of capability. Small moves reduced downside while accelerating learning.

Peter Sims, a Silicon Valley entrepreneur and Stanford educator, studied this exact approach across an unusually wide range of creative and business innovators, from Chris Rock refining comedy material in small clubs, to Howard Schultz testing the original Starbucks store concept in a single Seattle location, to the animation teams at Pixar iterating through thousands of story variations before committing to a final script.

In his book *Little Bets*, Sims argues that what looks in retrospect like visionary genius almost always turns out to have been built through a series of small, low-risk experiments, what he calls "concrete actions taken to discover, test, and develop ideas that are achievable and affordable." The method is not planned to be followed by execution. It is action followed by learning, repeated in rapid cycles.

The Chris Rock example really emphasizes how tangible the process feels. When preparing for a major performance, Rock doesn't simply sit at a desk refining polished routines. Instead, he goes to a small comedy club near his home to test rough, unfinished ideas in front of a live audience. He observes what works and what doesn't, then adjusts.

The following night, he tries again. Watching those early performances can be startling; they often seem painfully awkward. But that's intentional. Each failure offers valuable insights. Every audience reaction acts as a guide. Over weeks and months, Rock assembles a set from the fragments that withstand repeated testing.

The final show, perceived by the audience as effortless brilliance, was built from hundreds of small experiments, most of which failed. These failures weren't obstacles to the process. They were the process. Sims describes this as "failing quickly to learn fast," and the phrase captures an essential aspect of how direction works in practice. In uncertain environments, the cost of inaction, of waiting until you have enough information to guarantee the right move, almost always exceeds the cost of a small experiment that does not work.

A failed prototype teaches you something. A year of deliberation that ends without a decision teaches you nothing except that you are good at deliberating. Schultz did not commit to opening 1,000 Starbucks locations based on market forecasts. He opened one, watched what happened, adjusted the model, and expanded only after the experiment validated the concept. Jeff Bezos, as Sims documents, ran Amazon's early years as a series of small bets, testing categories, pricing models, and distribution methods, rather than committing to a grand retail strategy.

The same logic applies to career direction, especially in a period of technological disruption. When you cannot know in advance which path will pay off, which AI tool will become essential, which skill set will hold its value, which industry will be transformed next, the most productive strategy is not to deliberate longer but to run a small, affordable experiment and let the result teach you what deliberation cannot.

The "AI Moments" sessions I described above were exactly this kind of bet. They required no formal approval, no career restructuring, no public commitment. They were small enough that failure would have been invisible. But they generated real feedback, about what colleagues needed, about what the tools could do, about whether I had the capability and appetite to go further. That feedback was worth more than any amount of research conducted from the sidelines.

The second is to use sustained energy as a signal. In emerging fields, guarantees are rare. There is no reliable way to predict which specific role, tool, or skill will dominate in five years. What can be observed, however, is where energy consistently returns.

Energy appears through certain problems that catch attention longer than usual and topics that continue to interest you even after formal tasks are done. This leads you to read about them voluntarily, experiment without guidance, or think about improvements during unrelated times. This repeated attraction is important; if you stay interested even when it gets hard or confusing, that's worth noticing. If you are still willing to work on it after the initial excitement fades, it deserves further exploration. There was a period of about four months while writing this book when progress slowed significantly. What could have been completed in a few focused weeks stretched across an entire quarter. Work demands at Unilever intensified. Travel increased. Days were long, and evenings often left little mental capacity for research or writing. More difficult than the schedule was the doubt. There were multiple times where the argument stopped making sense, the structure kept changing, and the momentum that had carried earlier chapters slowed down or simply disappeared.

What did not change, however, was the underlying conviction that the theme explored in this book was something worth sharing. Even on weeks when little was written, the thinking continued. Notes accumulated. Ideas were revisited a hundred times. The energy was lower, but it was still present. That persistence was information. A passing interest would not have survived the combination of workload, travel, and self-doubt that accumulated along the way. The project stretched and slowed and became heavier than expected, but it did not stop, and that mattered more than early productivity ever could have.

Sustained pull toward something is its own kind of signal. Energy does not replace analysis, but it helps narrow choices when the future is unclear.

There is strong empirical support for treating this kind of persistent engagement as something more than anecdotal. Psychologists Richard Ryan and Edward Deci spent decades studying what drives human motivation, and their work on Self-Determination Theory, published in the American Psychologist, provides a rigorous framework for understanding why some pursuits hold our attention even in the face of difficulty while others fade at the first sign of friction. Their central finding is that intrinsic motivation, the kind that arises from genuine interest and personal significance rather than external rewards like money, status, or approval, is sustained by three core psychological needs: autonomy, competence, and relatedness. When an activity satisfies these needs, people engage with greater persistence, deeper creativity, and measurably higher well-being. When these needs are thwarted, even well-compensated work becomes draining.

Ryan and Deci define intrinsic motivation as "(...) *the inherent tendency to seek out novelty and challenges, to extend and exercise one's capacities, to explore, and to learn.*" What is worth noticing about this definition is how accurately it reflects the experience of pursuing a true professional interest despite challenges. It is not the kind of motivation that works like a New Year's resolution, arriving as a burst of enthusiasm that disappears once the initial excitement is gone. It is something more lasting than that: the tendency to keep returning to a problem even when it becomes hard, to keep exploring a domain even when the learning curve gets steeper, to keep building competence even when nobody is watching or rewarding the effort.

Their research also showed that the conditions surrounding an activity matter enormously. Tangible rewards, deadlines, surveillance, and pressured evaluations all tend to undermine intrinsic motivation by shifting the perceived source of behavior from internal to external.

On the other hand, environments that promote choice, recognize effort, and provide opportunities for self-direction are more likely to sustain engagement. This highlights how professionals should consider their direction during times of disruption. Choosing a path mainly because it looks good on a resume or because it was recommended by a colleague or mentor might create initial enthusiasm, but without a real internal motivation, such as a sense of autonomy and competence, that momentum is likely to decline under pressure. Conversely, selecting a path because one genuinely wants to learn about it and finds the problems engaging, despite the work being unglamorous, significantly increases the chances of enduring through the difficult middle stage when most people give up.

The energy that remained present through months of slow writing on this book, despite fatigue, competing obligations, and recurring doubt, was not discipline in the conventional sense. It was not willpower overriding reluctance. It was closer to what Ryan and Deci describe: an inherent pull toward exploration sustained by a sense of personal significance and growing competence.

In my own case, writing about AI adaptation was not assigned to me. Nobody was grading the output. I kept returning to the topic because it connected to a set of questions I genuinely found important, and because each attempt to articulate an idea sharpened my understanding of it.

That is the autonomy-competence loop at work in my own experience. Their research suggests that when this kind of energy persists even under adverse conditions, when a person keeps returning to a subject voluntarily, keeps investing discretionary time, and keeps thinking about improvements without being prompted, it is not random enthusiasm. It reflects a deep alignment between the activity and the person's core psychological needs. That alignment makes it a more reliable compass than most career frameworks or personality assessments could ever offer.

Three Simple Filters

When multiple options are available, and none come with guarantees, the temptation is to keep gathering information until one path clearly separates itself from the others. What helps more is a small set of filters applied to the options already in front of you.

The first filter is curiosity. Not the kind that looks good on a career plan or made sense a few years ago. The honest question is: what is pulling your attention right now?

A useful way to test this is to notice what you do when nothing requires you to do anything. For me, the answer kept showing up in the same place: opening ChatGPT throughout the day, not for work tasks, but to draw dogs as superheroes, sketch out business plans for companies that did not exist yet, or find better ways to raise my sons. No agenda, no audience. Just a genuine pull. That kind of repetition, unprompted and varied, is a different signal than a course you enrolled in months ago and never opened since. Curiosity sustains you through the parts that are slow, uncertain, and unglamorous.

The second filter is whether the direction can be tested before a full commitment is required. If the only way to explore a path is to quit a job, relocate, or make a decision that cannot easily be undone, that is worth pausing on. Strong directions almost always allow for a smaller initial entry point: a pilot project, a short course, or a prototype built over a weekend. My first serious experiment with AI did not require leaving Unilever or reinventing my career. It required opening a laptop and uploading a contract. The commitment grew only after the experiment produced something worth committing to. If no small version of a direction exists, the risk is probably disproportionate to what you know at that stage.

The third filter is whether the experiment will teach you something useful even if it does not work out. When I started generating images of dogs as superheroes using Nano Banana, there was no plan or intention. But that single experiment built familiarity with AI image tools, an understanding of what audiences respond to, and eventually a working knowledge of print-on-demand platforms when the idea turned into an actual small business. It could have gone nowhere, and I would still have walked away with something. If a direction builds capability or sharpens your thinking even in failure, the downside is limited. If it produces nothing unless it fully succeeds, it is closer to gambling than exploration. These filters do not eliminate uncertainty. They just make it more manageable.

Good Enough vs. Perfect Direction

Perfect direction does not exist in real time. It only looks inevitable in hindsight. What does exist is "good enough" direction, which is less about certainty and more about optionality. A good enough choice is one that opens more doors than it closes and creates learning that can be compounded.

In fast-changing environments, that bias toward learning and adaptability tends to outperform the search for a flawless plan.

A practical question follows. How does someone know when they have reached "good enough" and should stop researching? One signal is diminishing returns. When each new article, video, or course begins repeating what is already familiar, the activity still feels productive, but stops adding clarity. At that point, research can slowly turn into procrastination. Many capable professionals get stuck there, not because they are careless, but because preparation feels safer than committing to a move that might expose them.

That first ChatGPT test in late 2022, the one described at the opening of Chapter 1, was imperfect but practical. That practicality, even with clear limitations, made it obvious that the technology had crossed a threshold from interesting to usable. And yet, even after seeing that, the first response was familiar. For weeks, I did what many people do when they sense a change coming. I researched. I watched demonstrations. I collected articles. I enrolled in courses. I stayed busy, but little of it was going anywhere. It looked like progress without the risk of finding out whether any of it worked.

Progress started when experimentation replaced observation. I opened the tool frequently, applied it to small tasks, made mistakes, adjusted, and slowly built fluency through repetition. Over time, that practice moved from personal experimentation to real workflows, then to sharing outputs with colleagues, and eventually to leading internal workshops and joining Unilever's AI Champions network. Instead of readiness arriving as a feeling, it was built through usage.

A direction that is good enough to start with will teach you more in three months of movement than three years of deliberation ever could.

The Portfolio Approach

Direction carries so much weight because people tend to treat it like a single, irreversible bet. The perceived cost of being wrong feels high, which encourages over-analysis and delays movement.

What makes this harder is that the instinct to keep all options open, which seems like prudence, tends to work against satisfaction rather than for it. Psychologist Daniel Gilbert found in a series of experiments that people who made irreversible choices reported greater satisfaction than those who could still change their minds. We overestimate how much we need an exit and underestimate our own ability to commit to something and build from it.

The more resilient approach is to treat direction less like a single bet and more like a portfolio: several small experiments running at once, each designed to generate learning or open doors, even if none of them becomes the main path. This is how you build evidence before committing fully.

In practice, that might look like running internal workshops while also writing publicly, exploring an adjacent field, or prototyping something small with a new tool. Some experiments expand while others stall. The value is not that every initiative becomes permanent, but that each one builds skill, relationships, and options.

This matters particularly now because the environment increasingly rewards people who can connect across disciplines and adjust as tools and roles evolve. Direction becomes less about finding a fixed identity and more about building the kind of adaptability that holds up regardless of what changes next.

Adapter Profile: James, The Nurse Who Crossed into Health Informatics

James spent twelve years as an ER nurse in Atlanta. By his late thirties, the physical toll was becoming unsustainable. He wanted to stay in healthcare but could not see a clear path forward.

In 2024, his hospital piloted an AI-assisted triage system. Most nursing staff treated it as one more administrative burden. James treated it as a question: what would it mean if this worked? He started asking the implementation team how the system made decisions and gradually understood that most of the inefficiencies he dealt with every shift, the redundant documentation, the delayed lab results, the miscommunications between departments, were not clinical but information problems.

That realization gave him a direction. He enrolled in an online certificate program in health informatics, the field concerned with how data moves through healthcare systems, studying at night after work. He did not wait to finish before acting. He volunteered as the nursing liaison for the pilot, serving as a translator between clinical staff and the technical team. He became the person who could say: the algorithm is flagging this patient as low priority, but here is what it cannot see from the vitals alone.

Within eight months, the hospital's CIO asked James to join the informatics team full-time. Twelve years of bedside pattern recognition made him the most valuable person in a room full of data scientists. He was the one who knew what the numbers meant for the patient in the bed.

He did not plan any of that. He just followed the most interesting question available to him and remained useful for his organization at every step.

What Stops Us from Choosing

Not knowing enough is almost never the real reason people fail to act. More often, it comes down to fear, made worse by the belief that holding still is the safer option. Fear of losing what you have built, fear of how others will judge the decision, fear of starting over from a lower position: these are what make movement feel dangerous, even when doing nothing carries the larger risk.

The years already invested in a role, a company, or a professional identity can make any change feel like throwing something away. That pull is understandable but worth examining. Experience and knowledge do not lose their value when the context around them changes. They carry forward into new situations, often in ways that are not immediately obvious, and they tend to matter more in unfamiliar territory than in the environment where they were originally built.

Identity attachment can be even more restrictive. Titles become self-definitions, and self-definitions become limits. In environments where roles evolve quickly, identity cannot remain tied to a job label. It must be tied to capability, learning speed, and adaptability.

One antidote is to treat career moves as drafts rather than final versions. A move to California once seemed like the right direction based on an idealized image. The lived reality did not match the projection, and adjustment followed. The lesson was not that the original decision was unforgivable. The lesson was that speed of correction matters more than being perfectly right at the start. Professionals who thrive are not those who never choose imperfectly. They are those who notice, learn, and adjust quickly, before too much time is lost.

Direction Is a Practice, Not a Verdict

You do not "solve" direction once and move on. You choose, you test, you correct. Especially now, when the market moves faster than our plans.

If direction feels unclear, the smallest useful move is usually a short test of the thing you keep thinking about. The approach that worked for me is simple: pick one theme to explore and commit to a small daily habit of engagement. As psychologist K. Anders Ericsson showed, people build real capability through sustained, deliberate practice, not scattered attention. Fifteen to thirty minutes is enough if it happens consistently. As James Clear explains in *Atomic Habits*, lasting change comes less from big goals and more from small actions repeated until they become part of how you see yourself. Each completed session is a vote for a new identity, and those votes add up.

A loose weekly rhythm helps. The first week, observe and collect patterns. The second, apply what you are learning to one real task. The third is to share what you tried with one person, because writing it down forces a clarity that practice alone rarely produces, a finding supported by research on learning-by-teaching, including work by Aloysius Wei Lun Koh and colleagues. The fourth week, ask yourself two honest questions: did the work create energy or drain it, and did the learning curve feel alive or flat? If both answers point forward, run another cycle. If not, close the experiment and try something else.

A more detailed version of this 30-day experiment structure, with prompts and reflection templates, is available at adapterslab.ai/direction.

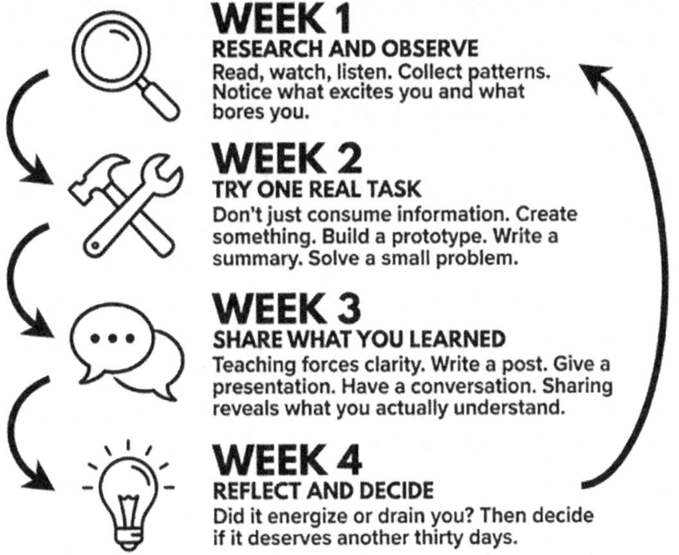

Figure 10 - The 30-Day Direction Experiment.
Made with Nano Banana PRO (2026).

Direction does not require a perfect plan. It requires a testable one. This experiment gives a theme four weeks to prove whether it deserves more of your time, and the structure is simple enough that a busy professional can run it without rearranging their life.

If you are waiting to feel 100% certain, you will be waiting while other people learn in motion. The better question is not "Is this the perfect path?" but "Is this a direction worth a two-week experiment?" Awareness shows you what is changing. Direction is deciding what you will try next. The next chapter is about the part most people avoid, that is, acting on before they feel ready.

7

ACTION

The biggest changes in my career happened when I moved before I had all the answers.

Not just thinking about AI but building with it. Not waiting for permission, but creating a mock-up, pitching an idea, testing a tool, or putting something small into motion before I had complete certainty. Those early actions created momentum, which changed how others saw me and how I saw myself. In my case, it meant moving from being seen as the person who knew about AI to being invited to lead larger, more visible projects.

What matters is that most of those turning points did not begin with a dramatic leap. As explored previously, they usually started with something small: a revised report, a prompt test, a short internal session, a prototype, a rough draft that was useful enough to start a conversation. You do not have to overhaul your life in one move. You just need to start moving in a way that produces evidence.

Awareness helps us see what is changing. Direction helps us decide where to go. Action is what moves us there. In the AI age, where tools evolve quickly and expectations change just as fast, the ability to experiment matters more than the ability to plan perfectly. This chapter is about learning through doing, building credibility through visible action, and closing the gap between knowing and doing before that gap becomes a career liability.

The Limits of Learning Without Action

Knowledge that goes unused becomes harder to apply when it matters. The longer someone waits, the more their understanding drifts from what the work requires today. At some point, knowing something and being able to do something with it are very different, and the market only pays for the latter.

This is where many professionals get lost. Inaction can look like productive busyness. It looks like reading every new article about AI without ever opening the tools already available to you. It looks like attending webinars, bookmarking courses, and taking pages of notes on digital transformation, while never applying a single idea to your actual workflow. It is the illusion of progress without any risk of failure.

In corporate life, this often hides behind respectable language. "We are still exploring options." "We will pilot when it is fully ready." "We need one more round of alignment." In reality, many people are waiting for certainty that never comes.

Jeffrey Pfeffer and Robert Sutton, both professors at Stanford, gave this problem a name. In their book The Knowing-Doing Gap, they studied why organizations and individuals consistently fail to act on knowledge they already have. Their conclusion was remarkable: the gap between knowing and doing is more damaging than the gap between ignorance and knowing. Some organizations do not fail because they lack good ideas. They fail because their culture rewards discussing, analyzing, and presenting those ideas more than it rewards acting on them.

Pfeffer and Sutton called this the "smart talk trap." In many organizations, the people who sound the most intelligent in meetings are the ones who identify problems, raise objections, and explain why something will not work. Sounding smart becomes a career skill, one that requires no action at all. The person who asks the sharp question during a strategy session gets credit for insight. The person who tries something imperfect and learns from the failure often gets less. This creates a culture where analysis is rewarded, and experimentation is quietly discouraged, not by policy, but by habit. The result is organizations full of knowledgeable people who are not doing much with what they know.

What makes this pattern especially relevant now is that AI has amplified both the volume of available knowledge and the speed at which it becomes outdated. You can learn everything published about generative AI in procurement today, and a meaningful portion of it will be different by the time you finish the reading. The problem is not a shortage of information. It is a surplus of information paired with a shortage of action. Pfeffer and Sutton found that the companies that closed this gap shared a common trait: they treated doing as the primary way of knowing.

They built cultures where trying something small and learning from it was more valued than producing a lengthy analysis of why something might not work. They judged people not only on what they knew but on what they had tried. That distinction runs through everything in this chapter. The colleague who accumulated certificates and the interns who built rough prototypes both had knowledge. The difference was who put it to use.

This is exactly what David Kolb argued in his work on experiential learning. His model describes learning as a cycle: concrete experience, reflective observation, abstract understanding, and active experimentation. In simple terms, we learn by doing something, thinking about what happened, making sense of it, and trying again with better judgment. Without that contact with real tasks, learning can stay conceptual for much longer than people realize.

More prominent research on experiential learning and deliberate practice has long pointed to this. In their landmark study, Ericsson et al. proposed that expert performance is not the product of passive exposure or innate talent but of prolonged, structured practice designed to improve specific aspects of performance through repetition, feedback, and successive refinement.

Courses can signal interest. Practice is what builds skill. I saw this clearly in a colleague from the Planning department who became genuinely interested in digital transformation. He completed multiple online courses, collected certificates, and kept excellent notes. His profile looked strong. He could speak fluently about the automation of planning systems with AI and emerging tools. But whenever it was time to apply any of that knowledge, he hesitated.

He was always still researching the best use cases, as if one more course would finally make starting feel less daunting.

At the same time, a group of interns was experimenting with tools already approved within the company. They built small dashboards, tested AI-assisted summaries, and created quick prototypes that solved narrow problems. None of it was perfect. None of it was especially polished. But it was visible, useful, and repeatable, making them gain visibility within our AI Champions group. Within a year, those analysts were leading pilots. The colleague who stayed in learning mode gradually lost relevance, not because he lacked intelligence, but because he kept postponing the moment of learning where others could see. That pattern matters. In times of rapid change, the people who adapt are not always the ones who know the most. They are often the ones willing to test, adjust, and learn while the work is still unfinished.

How Action Turns Learning Into Capability

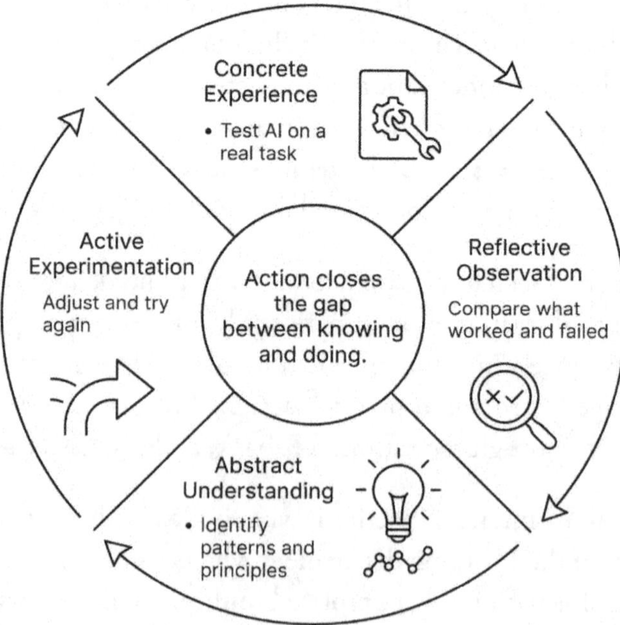

Figure 11 - How Actions Turn Learning into Capability. Made with Nano Banana PRO (2026).

The distance between learning and capability closes through contact with real work, not more study. The colleague who studied and the interns who experimented both invested time. The difference was where the learning became visible.

Why Small Action Beats Long Preparation

The instinct to prepare before acting is natural, but the feeling of readiness tends not to arrive on schedule. In fast-moving fields, the situation usually changes before the preparation catches up. Inside large organizations, I have watched teams spend months refining presentations for pilots that could have been tested in a week. As Eric Ries argued in *The Lean Startup*, the most effective way to learn is not through elaborate planning but through small, rapid experiments that put ideas in front of real users as quickly as possible.

His build-measure-learn cycle shows that a small prototype built early often teaches more than a polished business case that never leaves the slide deck. This is a pattern worth naming, and one I have fallen into myself.

In early 2023, before I began running internal AI workshops, I spent weeks researching digital adoption frameworks. I read consulting reports, watched talks, and compiled pages of notes. It felt productive. It also kept me conveniently distant from real exposure.

Nothing moved until I hosted the first short pilot session with a single colleague, using the tools already available at work. That small action generated clearer feedback than all the preparation combined. I could see what resonated, what confused people, what questions emerged, and what needed simplification. That first session did not prove readiness. It created readiness.

The study gives vocabulary. Action gives understanding.

Meaningful learning jumps almost always begin the same way: a small test before full confidence arrives. The late-night ChatGPT experiment I described in Chapter 1, the contract review that flagged a renewal clause I would have missed, was exactly that kind of test. The output was imperfect. It made a few confident assumptions and missed important context. But that single test captured the real lesson, that AI can be valuable and wrong at the same time, and the only way to learn where it helps is to use it on real materials. The shift was not theoretical. It changed how I approached the work, with AI serving as support for judgment rather than a concept to debate.

Early Experiments at Unilever

When Copilot became available internally, I applied the same logic safely within the company's environment.

I started with simple, low-risk tests. I used Copilot to rewrite supplier communications for tone and clarity. I ran cost variance scenarios in Excel and then asked Copilot to interpret the results. I drafted project summaries from Teams meeting transcripts. These were not high-profile use cases, but that was exactly why they mattered. They were ordinary tasks that occurred frequently, making them ideal for learning.

Each use taught me something that theory alone could not. I learned how much prompt clarity affects output. I saw how company guardrails shape what a tool can and cannot do. I also saw how much productivity can be unlocked through small automations that save time during a regular workday. My goal was never to master AI. It was to find out whether it could make my actual work easier. That kind of practical curiosity is what kept me experimenting, because it made the learning relevant and repeatable.

The first stage was informal. I would show colleagues how Copilot simplified tasks they already found frustrating. A summary that was faster to draft. An email that sounded clearer. An interpretation of data that cut through confusion more quickly. Curiosity spread, and before long, I was being asked to present on how procurement might use these tools more intentionally.

What made this difficult was not the technology. It was stepping forward without formal authority or a long list of credentials, presenting ideas that were still being tested in a culture that often values polish and certainty. But I had something better than credentials at that moment: I had results. I could show what worked because I tested it in the real flow of work.

That is where action starts to change status. A study can create credentials. Action creates authority. In the AI age, authority often matters more because people trust the person who can show what works, not just the person who can describe it well.

The Shark Tank Pitch

Winning that internal Procurement competition at Unilever mattered less than what the process exposed me to: executives who asked hard questions and expected clear answers, not enthusiasm. They do not want technical jargon. They want to understand the business impact, what changes it entails, why it matters, and what value it creates. That kind of clarity is hard to reach in isolation. It sharpens when you are forced to communicate under real conditions.

During the pitch, a live Copilot demo froze on screen. I acknowledged it, joked about the beauty of real testing, and kept going. That moment mattered more than a perfect run-through might have, because it was honest. The room saw the tool in action, including its imperfections, and that made the case more credible, not less. Credibility does not require perfection. It requires reality.

The Negotiation Practice Tool

One of the clearest examples of action outpacing process in my own work came when I was working as an AI Product Owner and had the idea for a Negotiation Practice tool.

The idea was ambitious. I wanted employees to be able to enter a simulated environment and practice different negotiation scenarios with AI. In the white paper, I outlined a concept centered on smart negotiation avatars, each representing a distinct negotiation style. These personas would not simply answer questions. They would role-play realistic dynamics, coach the user through the exchange, test the user's ability under pressure, provide a score, and document both strengths and areas for improvement.

It was a strong concept, but a complex product. At the time, I did not have a team of developers, designers, and engineers available to build it from the ground up. The budget was limited. Technical resources were already focused on other priority projects. If I had wanted a reason to delay the idea, I had several. Instead of waiting for ideal conditions, I decided to build a proof-of-concept myself.

Rather than beginning with the full traditional process of stakeholder meetings, repeated design sessions, long brainstorming cycles, and extended technical scoping, I created a mock-up version of the product using Manus. I did it carefully and without sharing any sensitive company information, using generic terms in the prompts, such as "the tool must be suitable to be used by buyers at a large consumer goods company." I also gave Manus the concept, the rationale, the desired learning experience, and the instructions for how the negotiation personas should behave, respond, coach, and challenge the user.

That still required serious work on my part over the course of a month. I had to research negotiation styles, define the personas, think through how each one should react in different situations, and shape an experience that felt realistic enough to be useful. Once that foundation was clear, Manus helped me turn the idea into something tangible: a functional and impressive proof of concept that made the product visible in a way a document alone never could.

That changed the conversation because I was no longer asking people to imagine the product. I was showing them a development-ready prototype. This was strategic for several reasons. It forced me to think through the product deeply enough that engineers and developers could clearly understand the experience I was trying to create. It gave me speed at a moment when speed mattered, because there was an internal opportunity tied to Unilever's annual Negotiation World Cup. It also reduced both the time and the cost required to get the concept off the ground.

If I had followed the normal route from the beginning, the product could easily have taken six months or more just to reach an early build stage, with significantly higher costs because so many people would have been pulled into the process before anything usable existed. By acting early and building the first proof of concept myself, I moved from idea to a credible prototype in under 90 days and at a fraction of the cost a traditional start would likely have required.

The important point is that the action was not a reckless choice. It was the strategic one. I had plenty of reasons not to do it: limited resources, limited budget, a technically demanding concept, and no dedicated team available at the start. But I could also see the opportunity. The momentum around the Negotiation World Cup created urgency, and urgency made speed valuable. By acting while that window was open, I created something concrete that others could understand, support, and continue building inside the company's approved boundaries.

This is often how progress works in practice. You do not always need the full machine running before you begin. Sometimes the smartest move is to build enough to make the vision visible. Once people can see the value clearly, support becomes much easier to earn.

Adapter Profile: Amara, The Small Business Owner Who Built an App

Amara runs a specialty spice shop in Houston. By 2024, her most common customer interaction was some version of: "I am making this dish. What spices do I need?" She answered that question dozens of times a day. Each answer drew on years of food science training. She loved giving the advice, but it consumed hours she needed for sourcing and blending.

She had heard about no-code app builders but assumed they were for tech people. Then a friend showed her the Lovable app. She decided to run a modest experiment to build a tool that lets customers enter a dish name and get a list of matching spices, with quantities and substitutions.

The first version was rough. Over two weeks, she refined the prompts and added her own blending notes. She embedded it on her website. Online orders increased by 30% in the first month because the tool recommended specific products from her inventory. In-store, she printed QR codes linking to the tool, freeing her to focus on customers who wanted deeper conversations.

Amara's story dismantles two assumptions: that AI tools are only for large organizations, and that acting on AI requires a detailed plan. Her plan was one evening, one tool, one specific problem to address.

As she told me, "I did not become a tech person. I just stopped assuming that technology was someone else's job."

Safe-to-Fail Experiments

The common thread across these examples is not confidence, but design. Instead of asking for the perfect plan, the more useful question is: what is the smallest step that produces new information?

A safe-to-fail experiment is small enough that failure costs little and learning arrives quickly. It usually has three characteristics. The downside is low, meaning the loss is limited mostly to time and attention, not credibility or career capital. The feedback is fast, meaning results show up in days or weeks, not quarters. The learning is transferable, meaning that even when the outcome is weak, it still improves the next attempt.

Many professionals skip this stage because they want to go straight to the big move. They want to launch the initiative, lead the transformation, or claim the title. The wiser path is to test whether the work fits, whether the value is real, and whether the skill can be built through repetition before scaling expectations.

The Champions Network at Unilever, which I described in Chapter 3, followed the same principle. I did not wait for a formal mandate to start something large. I started small enough that permission was not needed: a few colleagues, short sessions, one relatable use case per role, feedback after each meeting, and adjustments from there. Within a month, more than 100 employees had joined. Once it worked, support followed, and momentum replaced bureaucracy.

Before designing a full course, I ran a fifteen-minute demo using Copilot for meeting summaries. That quick test showed me what resonated, what confused people, and what needed simplification. Before applying for the AI Product Owner role described in Chapter 3, I volunteered on a short digital project. That experience gave me practical evidence that the role fit, without requiring a major leap, before I had tested the direction. Each of these actions was reversible. Each one taught me something. Each one opened a door that would have remained invisible if I had waited for certainty first.

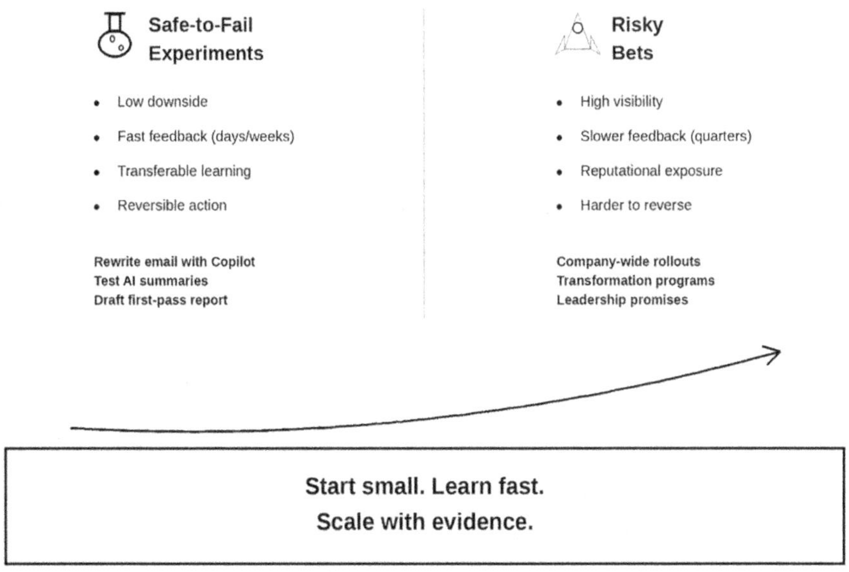

Figure 12 - Safe to Fail Experiments. Made with Nano Banana PRO (2026).

The difference between a safe-to-fail experiment and a risky bet is not ambition but design. Safe-to-fail experiments are designed to generate learning quickly, regardless of the outcome. Risky bets are stronger decisions when they come after smaller tests have already created evidence.

The Gap Between Knowing and Doing

Doing something real creates exposure, and exposure means accepting outcomes that cannot be controlled in advance. That is what makes action difficult, rarely the complexity of the steps themselves.

Professionals sense this without naming it. They add one more slide to the deck, take one more course, wait for one more signal that the timing is right. The language around it sounds responsible: I need to be ready, I need more training, I want to avoid getting it wrong in front of the wrong people. But underneath that language is a simpler fear: looking incompetent in front of others before the work is good enough to speak for itself.

The problem is that competence does not arrive before visibility. It develops through use, feedback, and repetition, which means learning is at least partly visible by nature. I felt this before the first Copilot workshop described in Chapter 3. The delay had nothing to do with the technology. I kept refining slides and rehearsing how the session might go, worrying about how it would land with peers. Once people were in the room, the fear dropped almost immediately. The anticipation of judgment was significantly worse than the event itself.

This fear is not irrational. It reflects how workplaces actually operate. Amy Edmondson, a professor at Harvard Business School, has spent more than two decades studying what she calls psychological safety: the shared belief within a team that it is safe to take interpersonal risks, speak up, admit mistakes, and try new things without being punished or humiliated. Her research, which began with a study of 51 work teams in a manufacturing company and has since been applied across industries, found that psychological safety is the strongest predictor of whether a team will actually engage in the behaviors that drive learning: asking questions, seeking feedback, experimenting, and discussing errors openly rather than hiding them.

What Edmondson found was not that some teams were braver than others by nature. It was the environment that shaped the behavior. In teams where leaders modeled openness, where admitting uncertainty was treated as a sign of honesty rather than weakness, and where failed experiments were discussed as opportunities for learning rather than liabilities, people were far more likely to act on what they knew. In teams where mistakes were punished or where looking polished mattered more than being honest, people held back, even when they had useful knowledge to contribute. The gap between knowing and doing, in other words, is often not a personal failing. It is a design problem. The conditions around people determine whether knowledge remains theoretical or becomes action.

This matters for anyone trying to experiment with AI inside a corporate environment. If the culture expects polish before people share work, if mistakes are career risks rather than learning moments, then even motivated individuals will default to preparation over action. Edmondson's research suggests that one of the most productive things a leader can do is to lower the cost of trying.

That can be as simple as running a short pilot with imperfect results or sharing your early, imperfect attempts with a tool before asking others to do the same. When the environment makes it safe to be a beginner, more people act on what they know. When it does not, knowledge stays locked behind the fear of being seen learning in public.

Formal training has real value, particularly around governance, responsible use, and quality standards. However, considering it a requirement before starting any experimentation reverses the effective sequence. By now, the idea of starting with small, low-risk experiments should sound familiar. It keeps coming up because it works. The difference here is what comes after, using training to build on what those experiments revealed. Training lands better on top of real experience than in place of it.

Starting over is also harder later in a career, and that deserves acknowledgment. Learning AI without a technical background can be quite humbling. The temptation is to wait until confidence arrives before beginning. Confidence does not tend to arrive first. It follows repetition, and repetition begins while confidence is still low.

This pattern aligns with what psychologist Albert Bandura described in his theory of self-efficacy: among the four sources that shape a person's belief in their own capability, mastery experiences, the direct evidence gained from performing a task, are consistently the most powerful. Observation, encouragement, and emotional readiness all play a role, but none substitutes for the evidence that doing provides. The real skill is not avoiding exposure as a beginner. It is moving through it quickly, repeatedly, and without treating that vulnerability as a reason to stop.

This will not happen only once. New tools will keep arriving, and each one will require a version of starting again. The professionals who adapt best are not those who avoid being beginners. They are those who have accepted that being a beginner again is a recurring part of staying relevant.

From Experiment to Identity

What ultimately changes is not just capability but identity. Every meaningful switch in my career, from procurement to sustainability to AI product work, began the same way: acting before feeling ready. I did not become an AI professional by studying AI. I became one by using the tools, building prototypes, teaching others, and solving real problems. The identity followed the action, not the other way around.

This matters because it changes how confidence works. People tend to assume they act because they are confident. More often, confidence grows because they act. Small experiments produce evidence, and evidence changes what others see in you. Over time, it also changes what you believe you are capable of.

James Clear, in his book Atomic Habits, offers a framework that explains why this sequence works. Clear argues that people tend to approach change from the outside in: they start with an outcome they want ("I want to lead an AI project"), then try to build the process to get there ("I should take a course on machine learning"), and hope the identity will follow ("Then I will be seen as an AI person"). The problem with this approach is that behavior driven solely by outcomes tends to fade once the outcome is achieved or motivation dips. Clear proposes reversing the sequence entirely. Start with the identity. Decide who you want to become, then let each small action serve as a vote for that identity.

Identity is the compound interest of behavior. It isn't a fixed state of being, but a *"repeated beingness."* Every action you take is a vote for the person you wish to become. One run doesn't make you a runner, but a thousand runs make the label a fact. You do not find your identity; you accumulate it.

This is what happened in my own transition. I did not sit down one day and decide to be an AI professional. I ran a Copilot demo for a colleague. I built a rough prototype on the weekend. I wrote a white paper. I volunteered for a digital project. I hosted a short workshop that went reasonably well. None of those individual actions felt like an identity shift at the time. But looking back, each one was a vote. Each one made the next step feel slightly more natural, slightly more like something a person like me would do. The identity did not precede the behavior. The behavior, repeated often enough, created the identity.

Clear's framework also explains why the fear of starting never fully disappears. If identity comes from repeated action, then every time a new tool or a new field emerges, you must begin casting votes again from scratch. That is a bit uneasy, but it also means the process is familiar to anyone who has done it before. You already know the sequence: act before you feel ready, notice what changes, and let the repetition build both the skill and the belief that you belong in the new space. The discomfort of being a beginner is not a sign that something is wrong. It is the cost of entry, and it gets easier to pay with practice.

One habit that reinforced this for me was keeping a short running note of what I tried and what happened. Not a formal journal. A dated list with one or two sentences per entry. Over time, the entries revealed patterns I would not have noticed otherwise: which tasks AI handled well, where it introduced risk, which prompts consistently produced better output, and where human judgment remained essential. When managers asked about the impact, the results were documented rather than vaguely remembered. That simple habit turned scattered experiments into a visible record of growing capability.

The experiments that lasted were the ones I told someone about. Explaining what happened forced me to understand what I had learned, and that made the difference more than any sense of accountability.

Sharing created a second loop: try something, tell one person what happened, and the next experiment almost always followed. That rhythm did more than motivation ever could.

If you want a structured way to run a week of focused experimentation, testing AI on real tasks, comparing outputs, teaching a colleague, and reflecting on what changed, I maintain a step-by-step version at adapterslab.ai/action. But the real version is simpler than any structure: open the tool you already have, apply it to one task you already do, and notice what changes.

Awareness shows what is changing. Direction helps choose where to go. Action is what moves you there. Not the perfect action, not the fully prepared version, not the one you will take once the conditions settle. The one available to you this week, applied to one real task, using the tools already in front of you.

8

POSITIONING

It is Not What You Do. It is How Others See What You Do.

If the right people cannot see what you are doing, it does not matter that you are doing it. Positioning is what makes your work legible. It is how people come to understand what you are good at, what you are building toward, and why they should trust you with a bigger scope.

I once worked with a colleague who consistently delivered. She solved complex supplier issues, rescued struggling projects, and kept budgets under control. The impact was real. The problem was that her work lived in spreadsheets and email threads, and she never spoke about it. Promotion season arrived, and her name barely surfaced. Not because leaders doubted her capability, but because she was hard to place. Her achievements were scattered across projects, absorbed into "business as usual," and easy to forget once the crisis was solved.

It is a frustrating lesson: being essential does not automatically translate into being recognized. As Jeffrey Pfeffer argued in his research on organizational power, the relationship between job performance and career outcomes, such as promotion or compensation, is positive but far from decisive. Doing good work is necessary, but it is not sufficient. Many professionals end up in this position. They become dependable engines of execution, while visibility flows to those whose work is easier to summarize, repeat, and associate with a clear theme.

Dorie Clark, a strategy professor at Duke University's Fuqua School of Business, spent years studying how professionals build lasting recognition. In her book The Long Game, she observed that becoming known as an expert in any field follows a pattern that most people underestimate. Based on her research and work with hundreds of professionals through her Recognized Expert community, Clark found that it typically takes two to three years of consistent effort before the first signs of traction appear, and closer to five years before others begin to recognize you as a genuine authority in your space. That timeline surprises people who expect faster results, and it explains why so many give up too early. The ones who persist, Clark argues, benefit from what she calls a compounding effect: once your name becomes associated with a specific area of value, opportunities begin to arrive without you having to chase each one individually.

Clark's framework centers on what she calls content creation as a visibility engine. The idea is about giving others a consistent way to discover your expertise. Writing a short internal summary after a project, sharing a lesson learned in a team meeting, or publishing a practical insight on a platform like LinkedIn all serve the same function: they create entry points for people who do not yet know your work.

Over time, those entry points accumulate. They make it easier for someone to explain what you do to a third person, which is one of the strongest indicators that positioning is working. Clark also emphasizes the importance of strategic patience. She argues that professionals should maintain no more than two or three primary goals at any given time to keep their efforts focused rather than scattered across too many fronts. When the work is focused, and the communication around it is consistent, the reputation builds on itself. That is the compounding Clark describes, and it is the same pattern I saw play out in my own trajectory with AI.

What makes Clark's perspective especially relevant is that she is not talking about personal branding in the superficial, influencer-driven sense. She is describing the slow, deliberate work of becoming the person others think of when a particular challenge comes up. In a corporate setting, this means that the colleague who is always associated with a clear contribution, whether it is process improvement, data quality, or cross-functional coordination, will be top of mind when a new initiative launches.

The one whose work is scattered and unnarrated, no matter how valuable, will be harder to recall. Clark puts it simply: the professionals who play the long game do not chase every trend or pivot with every news cycle. They pick a direction, stay close to it, and let the evidence build over time. That kind of discipline is exactly what separates positioning from noise.

I saw this dynamic firsthand. When I started sharing AI experiments at Unilever, there were months where the response was minimal. A post might get a few reactions, or a workshop might draw a handful of attendees. It would have been easy to interpret that silence as disinterest and stop. But Clark's research explains why persistence matters so much in those early stages. The compounding effect does not happen overnight. It happens after enough people have encountered your name in enough contexts that it begins to stick. For me, the shift came somewhere around the eighteen-month mark. Colleagues who had ignored my earlier posts started reaching out with questions. Managers who had not attended my workshops were now referencing them in meetings. The work had not changed, but the accumulated visibility had reached a threshold at which it began to generate its own momentum. That is the long game Clark describes, and it is one of the most underappreciated dynamics in career building.

For a long time, I assumed results were the simple recipe for success. I thought competence would be obvious, and that the right leadership would connect the dots between what I was contributing and the opportunities I deserved to be considered for. In a large organization, good work does not speak for itself, and the people who could recognize it are almost always too absorbed in their own priorities to notice it.

Positioning is not limited to public platforms or LinkedIn posts. It happens in smaller moments that almost never get labeled as personal branding, even though they shape careers far more than any single post does: how you introduce yourself in meetings, what you highlight in updates, how you summarize a win, what you share when someone asks what you are working on, what your profile signals when a colleague looks you up before a call. Over time, those moments form a picture that other people carry around of you, and that picture influences what you get asked to do next.

When I began sharing my AI journey on LinkedIn, in internal talks, and through everyday project communication within the company, people started to understand my work through a different lens. The work itself had not fundamentally changed, but the way it was perceived had. Others could now describe it clearly and, just as importantly, repeat that description to someone else.

The Line Between Sharing and Selling

A typical response is to withdraw from the concept of visibility because it seems like self-promotion. I had the same response. I associated visibility with arrogance and noise. Over time, I realized I was mixing up two very different behaviors. One is performative. The other is clarifying.

Many professionals avoid sharing their work because they fear sounding arrogant. I used to resist the idea of self-promotion, associating it with people who were louder than they were capable and treating visibility as something that belonged to them rather than to the work itself. But I was wrong about what positioning really means.

Authentic positioning means making your value clear and accessible, helping others understand what you do and why it matters, without waiting for them to figure it out on their own. The difference between authentic positioning and hype is simple: authenticity shows your process and results. Hype makes claims without proof. The key is to tell stories that show real outcomes and learning moments. Not "I am an expert." But "Here is what I tested, here is what surprised me, here is what worked."

When I first started talking about AI internally, I was not showing off. I shared what I was learning in plain language to help colleagues who were as confused as I was. That transparency built trust. I was not the AI guy because I claimed it. I became that because I consistently showed it.

There is growing evidence that this kind of authentic visibility produces measurable career benefits. A large-scale study published in Frontiers in Psychology, led by researcher Sergey Gorbatov and colleagues, examined the effects of personal branding on career outcomes across 477 professionals in both Western and Asian work environments. The researchers developed a validated measurement scale for personal branding and tested its relationship with employability and career satisfaction. Their findings were clear: professionals who actively engaged in personal branding reported significantly higher perceived employability, and that employability in turn led to greater career satisfaction. The statistical relationship was strong. Personal branding explained 43 percent of the variance in career satisfaction when employability was included as a mediating factor. In plain terms, the people who made their work visible and consistently communicated their professional value felt more in control of their careers and more satisfied with where those careers were heading.

These findings are especially useful because the researchers distinguished between the intention to engage in personal branding and the actual practice of doing it. Many professionals think about positioning but do not follow through. The study showed that career achievement aspiration was the strongest predictor of whether someone acted. People who had a clear sense of where they wanted their careers to go were far more likely to engage in the consistent, visible communication that drives results. Career self-efficacy also mattered: the belief that you can execute career-related behaviors well made it more likely you would do the work of positioning yourself. This is not about personality type or being naturally outgoing. It is about setting a goal and aligning your communication habits with it.

The study also highlighted a facet of personal branding that is often overlooked in conversations about self-promotion: it is a signaling function. When you communicate your professional value clearly, you reduce what economists call information asymmetry. Decision-makers inside organizations and across industries have limited time and attention. They cannot evaluate everyone's contributions firsthand. What they can do is notice patterns in who consistently shows up with clear, useful, and credible messages about their work. Gorbatov and colleagues found that this signaling was positively associated with career success across multiple dimensions, including social capital, financial rewards, and career opportunities. The professionals who were most visible were not necessarily the loudest. They were the ones who had built a clear, honest, and repeatable message about what they could do and what they had already done. That is the version of personal branding that matters in a business context, and it aligns with every lesson I learned inside a large organization.

One detail from the research that stood out to me was the finding about career feedback. Gorbatov and colleagues found that receiving career feedback was negatively related to the intention to engage in personal branding. In other words, when professionals already receive clear direction from managers or mentors about their career prospects, they feel less urgency to position themselves. The problem, as anyone in a large organization can attest, is that professionals rarely receive meaningful career feedback. Performance reviews tend to be too slow, too generic, or too focused on the past to provide forward-looking guidance. In the absence of that guidance, personal branding fills the gap. It becomes a way for professionals to take ownership of how they are perceived rather than waiting for a leader to define it for them. That is not self-promotion for its own sake. It is a practical response to how organizations actually work.

A simple test works in almost any context: are the words pointing to proof, or asking people to believe?

Authentic Positioning vs. Empty Hype

AND HOW TO UPGRADE IT

AUTHENTIC SIGNAL	EMPTY HYPE	UPGRADE MOVE
Shows results	Makes claims	Add one concrete outcome — time saved, errors reduced, cycle time improved.
Uses examples	Uses slogans	Swap a tagline for a 2-sentence micro-story: context → action → result.
Admits learning	Pretends mastery	Name what is being tested and what changed after one iteration.
Focuses on value to others	Focuses on self-image	Finish with the "so what" for the team, customer, or decision-maker.
Invites dialogue	Demands attention	Ask one specific question that encourages informed responses.
Explains the work simply	Leans on buzzwords	Replace trendy terms with plain language and one real task it improves.

Authentic signals are observable, specific, and audience-centered.

Figure 13 - Authentic Positioning vs. Empty Hype. Made with Nano Banana PRO (2026).

Positioning done right does not make you louder. It makes it clearer. The difference between the two columns is not volume. It is evidence.

Nobody wakes up trying to sound hype driven. It usually shows up when the work is real, but the language remains abstract. The fix is usually a small detail, an example, a measurable change in the process, or even a sentence about what was learned after one round of testing.

You can hear the difference in common phrases. "Passionate about AI transformation and innovation" is too vague to be useful because it gives people nothing to hold on to. A sentence that describes the actual work is far more credible: "Built an AI-assisted workflow that cut monthly reporting from six hours to forty-five minutes, then shared the template so the team could reuse it."

The same goes for "Expert in prompting and AI strategy." It reads stronger when it describes actual practice: Testing prompts inside real workflows. The current approach improves first-draft quality and reduces rework, and the prompt library is updated each week based on what fails.

Even leadership language improves when it becomes concrete. Driving cross-functional alignment and change" gets clearer when it names the action: "Ran a two-week pilot with Legal, Finance, and Ops to standardize intake. The new process removed duplicate requests and shortened review cycles.

Good positioning does not require more volume. It requires clearer proof.

Making Positioning Work

The shift from running small AI experiments to being introduced as the internal AI lead did not happen because a new identity was announced. It happened because the work became visible and easy for others to describe. People could point to something tangible and explain it to others without embellishing it. That is how reputations form inside organizations.

The reputation came gradually, not from a single defining moment but from a series of small ones that accumulated into a pattern others could recognize and repeat. A significant part of that came down to language. Describing the work as tool exploration generated little response. Describing it as finding ways for procurement teams to work more effectively with AI opened different conversations, because it framed the work as a business contribution rather than a personal interest.

As the workshops grew into the AI Champions network I described earlier, something else became clear. Moving into a space without formal credentials or established authority behind you is unsettling in ways hard to explain to those who have not done it. The way through it was simple: I stayed close to what I could honestly claim, which was applied learning. I shared what I was doing, what was working, and the limitations.

Inside Unilever, I framed AI as a practical enabler. I talked about decision-making, cycle time, and quality. I also used short case snapshots in emails and meetings, so the progress was obvious. For example, one workflow reduced report preparation time by about 60 percent. Numbers like that helped, but I noticed something else mattered almost as much: colleagues could see a repeatable method, not a one-off trick.

Different audiences read positioning through different lenses, and it is worth being aware of which one you are addressing. Within organizations, people look for trust and usefulness. They care about whether someone can simplify decisions, remove obstacles, and help others move faster. Managers scan for readiness and reliability. Senior leaders focus on business impact and risk reduction. Outside organizations, people look for consistency: what someone is becoming known for and what they repeatedly show up to do. Building a strong internal reputation is the more immediate priority. External visibility can follow from that, but it does not have to come first.

My own approach was to define a through-line I could stick to. For me, it was transformation and learning. Procurement, Sustainability, AI: different domains, same pattern. I was drawn to change and to helping teams adapt. Once I started repeating that theme, using the same few proof points across meetings, posts, and project updates, people stopped guessing what I "really" did. They could explain it.

Adapter Profile: David, The Auditor Who Made His AI Skills Visible

David had been a senior internal auditor at a Midwest bank for nine years. His work was precise, detail-oriented, and largely invisible.

In early 2025, he began using AI to accelerate audit workflows. He trained a custom GPT to review policies against regulatory standards, cutting preliminary review time by forty percent. He used Claude to draft executive summaries in language board members could act on. He built a simple AI-assisted dashboard that tracks open findings in real time, replacing a spreadsheet that three people manually updated every Friday.

The work was valuable. What changed his career was what he did next: he made it visible. He wrote a short internal case study with specific numbers: hours saved, error rates reduced, turnaround time compressed, and improved overall quality of the outcome. His manager forwarded it to the chief audit executive, who mentioned it in a board meeting. Within two weeks, David was presenting to senior leadership.

Then he wrote about it on LinkedIn. A practical post: the problem, the tools, the results. It reached over fifteen thousand views. A fintech company offered him a role as head of AI-augmented compliance, a position that had not existed six months earlier.

Positioning is translation: taking what you have learned and expressing it in terms others can understand, value, and act on.

Building a Portfolio That Speaks for You

I also began treating my career as a portfolio rather than a sequence of roles. Publicly, LinkedIn became a record of applied learning. Internally, I kept a running set of mini case snapshots using a simple structure: problem, action, result, lesson. Having those ready meant I did not rely on memory when an opportunity came up.

A portfolio does not have to be full of metrics to be credible. In many roles, the most persuasive proof is an artifact or a process change. A before-and-after draft. A template. A one-page guide. A prompt library. A decision that changed because the analysis became clearer. Adoption by others without being chased. A short endorsement from someone who benefited. Those are all forms of evidence people recognize.

This also changes the way credentials fit into the story. I did not have an AI or computer science degree. What mattered was whether the work improved outcomes and whether I could help others reproduce it.

The same logic applies to the past. Older work can support a new direction when you describe it in terms of the capabilities it has built. I reframed sustainability work from my master's program as an early exploration of decision models and responsible adoption patterns. That connection held because the underlying skill set carried over: systems thinking, measurement, governance, and making trade-offs under constraints.

This portfolio mindset has a close parallel in the work of Reid Hoffman and Ben Casnocha. In The Startup of You, Hoffman, the co-founder of LinkedIn, and Casnocha, a venture capitalist, argue that every professional should manage their career as an entrepreneur would manage a startup. The core idea is that careers are not static. They are living, evolving ventures that require the same adaptability, risk awareness, and investment mindset as any early-stage company.

Hoffman and Casnocha call this operating in "permanent beta," a term borrowed from the software world. It means treating yourself as a work in progress, never assuming your current skills or reputation are enough, and always looking for the next area of growth. For anyone building an AI-related career from a non-technical background, this framing is especially relevant. You are not pretending to be something you are not. You are running a series of experiments, learning from each, and letting the results guide your direction.

One of the most practical concepts in the book is the author's "ABZ planning." Plan A is what you are doing now, your current role, and the direction you are actively pursuing. Plan B is the pivot you are ready to make when new information suggests a better opportunity or when your current path loses momentum. Plan Z is your fallback, the safe ground you can return to if things go wrong. This framework is useful because it removes the pressure of getting career positioning exactly right the first time. Instead, it encourages iterative positioning: test a direction, gather evidence, adjust. That is exactly what a career portfolio does. Each project, each case study, each visible contribution becomes a data point that informs the next move. The professionals who adapt fastest are not the ones who planned everything in advance. They are the ones who built systems for learning, pivoting, and communicating what they learned along the way.

Hoffman and Casnocha also make a compelling case about the role of networks in career positioning. They argue that "the fastest way to change yourself is to hang out with people who are already the way you want to be." Your career success, they write, depends on both your personal capabilities and your network's ability to amplify them. They describe this as "I to the We," in which an individual's power is enhanced through a professional network. In practical terms, this means that positioning is not just about what you say about yourself. It is also about who can say it on your behalf.

The colleagues who forward your case study, the manager who mentions your name in a leadership meeting, the contact who recommends you for a role you did not know existed: these are all functions of a network built through consistent, visible, and valuable contributions. Hoffman describes three types of professional relationships that matter: allies who serve as close collaborators and advocates, weak ties who sit two or three degrees away and can open unexpected doors, and followers whose attention translates into career capital. Building and maintaining these connections is not separate from positioning. It is how positioning scales beyond your immediate circle.

If someone wants to tighten positioning quickly, it usually comes down to a few moves repeated consistently. The one I found most useful was asking a simple question: if a colleague described my work to someone who had never met me, would they get it right? If the answer is no, the story needs tightening, not the work. From there, the practical steps are small. Adjust one sentence that appears often, a headline, a bio, or a meeting intro, so it reflects direction as well as current role.

Keep one or two short case snapshots handy, written in the problem-action-result-lesson format. Share one when the moment fits, whether that is a manager update, a team channel, or a public post. Ask a trusted colleague if the story is clear and believable, then tighten the language. None of this requires a branding exercise. It requires paying attention to how you describe what you already do.

One more detail worth noting: visibility must fit reality. In corporate settings, confidentiality and internal politics are real constraints. There are safe ways to share. Methods and lessons usually travel without risk. Vendor names, sensitive numbers, or specific deal details often do not. Internal sharing is usually the easiest starting point and often the most immediately useful for career momentum.

Why the Resistance Is Real

Positioning is not difficult because the work is missing. It is difficult because drawing attention to it can seem like the riskier move. Imposter syndrome shows up early, especially in fast-moving spaces like AI. I felt it when I began posting and leading sessions. The anxiety did not go away because I convinced myself I was qualified. It eased as the work became more grounded and repeatable. Each experiment, each workshop, each improvement created another data point. Many professionals have been shaped by environments where drawing attention to your work was seen as overstepping, and where the expectation was to deliver and let others draw their own conclusions. The practical consequence of that conditioning is that silence tends to be misread. In most cases, it gets interpreted not as humility but as absence.

Skepticism is part of the terrain, too. When I leaned into AI more publicly, I could feel it in the background. I understood it. Workplaces have seen plenty of trends come and go. What helped was staying close to outcomes and repeatability. I shared small workflows that made work easier, small improvements that reduced rework, and small lessons that helped colleagues avoid mistakes. Over time, that built trust with the people who were open to it, which was enough.

Pfeffer makes a blunt version of this argument in *Power: Why Some People Have It, and Others Do Not*. Performance matters, but it does not automatically translate into opportunity. Being understood by the right people and being visible enough to be remembered affects who gets picked when the stakes rise.

Herminia Ibarra reached a similar conclusion in *Working Identity*. Career change does not come from thinking your way to an answer. It emerges through action, through new experiences, and through the effort to name what those experiences are becoming. People try on identities, adjust them, and discover what fits. This is not branding in the superficial sense. It is identity formation in public view, one iteration at a time.

Positioning Creates Opportunity

For years, I assumed the work would speak for itself. The results were real, the effort was there, and I expected the right people to connect the dots. I missed that careers do not move on effort alone. They move on what people understand about the value being created, and what they can reliably associate with a person over time.

Things changed when I started using clearer language to describe what I was already doing. I shared AI experiments on LinkedIn, used internal forums when they made sense, and framed project updates around what was changing in the workflow and why it mattered. The work did not suddenly become more impressive. It became easier to recognize. Over time, that recognition turned into a pattern: not "someone trying tools," but "someone building capability and applying it in a business context."

That is positioning. Not performance, not rebranding, and not pretending to be something else. It is the work of making sure the story matches the trajectory. Sometimes a few small changes are enough: a headline that reflects the direction of the work, a short case study that captures a problem, a method, and a result, or a clearer sentence in a meeting about what was learned and why it matters. These details may seem minor, but they accumulate.

Chapter 9 takes it from narrative to repeatability. Positioning helps with work travel, but tech empowerment makes the work stronger and easier to reproduce. The ADAPT method is the structure for doing that, so progress is not dependent on motivation or luck.

9

TECH EMPOWERMENT

Do Not Get Lost in the Tech. Make It Work for You.

I hit the wall sometime in 2023 when opening the browser to test a new AI writing tool. By the time it loaded, posts on LinkedIn were already talking about alternatives to that tool specifically. The feed was full of "must-try" apps, "essential" updates, and lists that promised the same thing: if you do not catch up, you will fall behind. So, the instinct was to do what many people do: save everything. Telling myself there would be time to come back later, once a slower week arrived. Weeks passed, and that folder kept growing. Opening it started to feel like pressure. Not learning. Just collecting options. And because of my involvement in AI adoption at work, the pressure felt worse, with a recurring thought: "If I am taking this seriously, shouldn't I know all of them?"

One evening, the pattern became impossible to ignore. Hours had gone into reading about new tools, saving links, and following trails across newsletters and forums, and almost none of that time had produced anything useful at work. The bookmarks folder was deleted that night. The approach changed with it: one new tool at a time, and only if it addressed something that was getting in the way that week. That single constraint did more to clarify the process than any course or framework had.

The constant stream of new releases did not slow down, but it lost its grip on how time got spent. The focus returned to something far more practical:

find a task that repeats, find a tool that makes it faster or better, keep what works, and move on from what does not. Progress began showing up in the work itself rather than in a list of things still to be tried, and that made it considerably easier to stop measuring how far behind the curve things were.

There is another reason that the approach works. People rarely get overwhelmed by too many tools. They get overwhelmed because every new tool looks like a decision they have to make. Once you reduce the number of active decisions, your attention comes back, and you can focus long enough to build a workflow that makes sense.

There is a name for what happens when options multiply faster than your ability to evaluate them. Barry Schwartz's Paradox of Choice, introduced earlier in this book, becomes even more relevant here. When options multiply faster than your ability to evaluate them, the result is not better decisions. It is anxiety, regret, and paralysis. People spend more energy evaluating than acting. They second-guess what they picked. They imagine better alternatives that they did not explore. And in many cases, they avoid choosing altogether, not because they lack motivation but because the cost of deciding feels higher than the cost of doing nothing. His research showed that people presented with fewer options were more likely to make a purchase, more confident in their choices, and more satisfied afterward than those given a wider selection.

The parallel to AI tools is hard to miss. Every week, new platforms arrive with overlapping features and competing claims. For someone trying to build a practical working relationship with AI, the market can feel more like an exam with no right answer than a menu. Schwartz offered a useful distinction between two types of decision-makers. Maximizers seek the absolute best option. They research exhaustively, compare endlessly, and often end up less happy even when they choose well, because they cannot stop wondering whether something better was out there. Satisficers, on the other hand, define what "good enough" looks like and stop searching once they find it. They are consistently more satisfied, less stressed, and faster to act. When I deleted that folder of bookmarks and committed to one tool at a time, I was making a shift from maximizing to satisficing, without having the language for it.

Schwartz also pointed out that the costs of choosing are not just emotional. They are cognitive. Every open decision occupies mental bandwidth that could be spent on the work itself. This is what researchers call decision fatigue: the more choices you process in a day, the worse your judgment becomes as the day goes on. In the context of AI adoption, this means that trying to evaluate ten tools at once does not make you more prepared. It makes you less capable of doing focused work with any of them.

The fix, as Schwartz suggested, is not to avoid new tools entirely. It is to reduce the number of active decisions at any given time. Set a filter. Apply it. And move forward with what passes, knowing that a good-enough tool used consistently will always outperform the perfect tool you never settled on. That principle became the backbone of how I managed my own AI toolkit, and it is the same advice I give to anyone who feels stuck in the browsing loop.

Schwartz published The Paradox of Choice in 2004, years before AI tools became part of everyday work. But his argument has only grown more relevant. The number of AI applications available today is orders of magnitude larger than the number of productivity tools that existed when he wrote the book. If anything, the paradox he described has intensified. The professionals who handle this effectively are not the ones who test every tool. They are the ones who have learned to ignore what appears in their feed, commit to one or two options that meet a clear need, and stay with those long enough to build real skill. That is a discipline, and one that can be learned.

In practice, applying a satisficing approach to AI tools means defining your criteria upfront, testing against a single real task, and going with the first option that meets them. The time you save by not endlessly comparing is time you reinvest in building actual fluency with the tool you chose. Over weeks and months, that fluency compounds into something far more valuable than having browsed a wider range of options.

One question I get when sharing my story of reinvention is whether I had a background in computer science. The answer is no, and that is precisely the point. In the AI era, it is easy to assume that thriving requires becoming technical. In practice, the bar is lower and more realistic

. You do not need to understand the code behind these systems. You need to get familiar with them by using them in ways that support your work. That is what I mean by tech empowerment. It is the ability to pick tools that fit your environment, learn them fast enough to get value, and keep your progress even when the apps, features, and headlines change.

This chapter is about building that kind of relationship with technology. Something steady. Something you can repeat.

A Pattern That Runs Through My Career

Looking back, the line from DOS to SAP to AI is consistent. Curiosity comes first. Structure follows.

At Securit, the DOS machine did not look approachable. Most people avoided it. I didn't have a background, but I was determined to learn. A colleague showed me Lotus 1-2-3, and suddenly I could automate calculations that used to take hours. I did not understand how the computer worked. I understood how to make it useful.

At DuPont, the SAP wave was bigger and more intense. The migration I described in Chapter 1, moving thousands of purchase orders from the old IBM system, meant long days working across both systems simultaneously. That effort led to something I did not expect. One of the procurement managers, Fabio, saw the contribution and invited me to join the implementation team launching DuPont's first SAP version in procurement in Brazil. At eighteen, I became an SAP key user.

That period shaped how I think about technology. I did not become a programmer or a systems engineer. I became someone who knew how to make the technology work for the task at hand. None of the transitions that followed, not the move to Unilever, not the entrepreneurial years, not the return to corporate life, required me to become technical in the engineering sense. What each one required was learning systems that could be implemented quickly enough to improve outcomes and reduce the effort required for routine work.

In 2022, that same pattern repeated with ChatGPT. The contract test described at the opening of this book delivered the same message in a new form: the value was practical, not theoretical. Fewer mistakes. Faster review. Better attention to the parts that required judgment. This lines up with Ethan Mollick's argument in *Co-Intelligence*. AI delivers the most value when it becomes a steady partner in work you already repeat, not a magic replacement for thinking.

By 2025, the curiosity moved from analysis into building. CampBear, made with the Lovable app, started as a family idea while planning a trip to Iceland. I wanted a private space where we could upload photos and short videos as we traveled, add notes in the moment, and later turn it into a clean, printable memory book. I had no coding background. I had a clear picture of what I wanted and the patience to iterate after watching a few short YouTube videos on working with Lovable. I would describe a feature in plain language, review the tool's output, and fine-tune the prompt until the flow made sense. It reinforced an important point: technology is becoming more accessible to people who can clearly express what they want and keep refining it until it works.

Adapter Profile: Mei, The Supply Chain Analyst Who Became an AI Trainer

Mei spent seven years as a supply chain analyst at an electronics manufacturer in Southern California. By 2024, her company was piloting AI-driven demand planning tools, and the first thing those tools did well was exactly the work that filled her calendar.

Rather than waiting for a reorganization, Mei built what she called her "parallel stack." Evenings and weekends, she learned how to use the same AI tools her company was evaluating, not just as a user, but as someone who could configure them, identify blind spots, and explain outputs to non-technical colleagues.

She used ChatGPT for exploratory analysis, Claude for comparing vendor documentation, Perplexity to track industry implementations, and built a personal knowledge base in Notion, organized by tool, use case, and likely colleague questions.

When her company launched the pilot, Mei was the most prepared person in the room. Not because she had a data science degree, but because she understood the gap between what the tools could do and what the organization needed. She was selected as the pilot's internal lead. Six months later, she was promoted to AI Implementation Specialist for Supply Chain.

Tech Empowerment means building practical fluency with a few tools that serve your specific context and bridging the gap between technology and the people who need to use it.

Cut One Task in Half

A practical method I relied on during my master's program and while writing this book starts with a simple constraint: one task, one change, one clear comparison. Not a full workflow overhaul, not a new system. Pick one task that repeats, apply AI to one part of it, and see whether the time or effort required reduces.

The easiest place to start is a task you already do regularly, find tedious, and would not miss if it took half the time. The one you postpone not because it is difficult, but because the setup cost feels disproportionate to what it produces.

Repetition is what makes the improvement visible. When you try to change too many things at once, you lose the ability to tell what helped. A single recurring task gives you a clean before-and-after without having to guess.

Writing is the clearest example I can offer because the baseline is easy to describe. Turning rough notes into a drafted subsection used to take around ninety minutes: rereading everything, finding the thread, outlining, drafting, and then rewriting for clarity. I started using AI for the first half of that process, only the structure, and a first-pass draft.

 I would paste raw notes and ask for three things in one prompt: an outline, a draft written in my tone, and a short list of gaps or weak logic worth addressing. The second half, the argument, the judgment, the final tone, was never handed over. What changed was how much effort the start required, which is often where most of the time goes anyway. This pattern is consistent with what Noy and Zhang found in a controlled experiment with 453 college-educated professionals: access to ChatGPT reduced average completion time by 40 percent and raised output quality by 18 percent, with the task moving away from rough drafting and toward idea generation and editing, the parts that require human judgment.

A larger-scale study led by Erik Brynjolfsson, Danielle Li, and Lindsey Raymond at Stanford and MIT confirmed this pattern using data from over 5,000 customer support agents at a Fortune 500 company. When those agents were given access to a generative AI assistant, their productivity increased by 15 percent on average, measured by the number of customer issues resolved per hour.

But the real finding was who benefited. The least experienced and lowest-performing workers saw gains of roughly 30 percent. New agents using the AI tool reached the same performance level in two months that untrained agents typically needed six months to achieve. Meanwhile, the highest performers saw only small improvements, and in some cases, no measurable change at all. The AI was not replacing skills. It was compressing the learning curve for the people who needed it most.

What makes this study especially useful for the argument in this chapter is that it shows what happens when AI is attached to a specific, repeatable task rather than introduced as a broad capability. The support agents were not asked to become AI experts. They were given a tool that suggested responses in real time, which they could accept, modify, or ignore. Those who followed the tool's recommendations closely improved the fastest.

The researchers also found that AI assistance changed how customers treated the agents: people were more polite, less likely to ask for a supervisor, and generally reported better interactions. Employee turnover among agents using the AI tool dropped by nearly 9%, largely driven by the retention of newer workers who were no longer overwhelmed by the challenges of their first months on the job.

The connection to what I experienced while writing this book is direct. When I used AI to handle the first pass of a draft, I was not handing over judgment. I was compressing the setup phase to get to the judgment faster. The agents in Brynjolfsson's study were doing the same thing in a customer service context.

The AI handled routine language, allowing the agent to focus on the customer's actual problem. In both cases, the benefit did not come from the technology being brilliant. It came from the technology reducing friction in a task that repeated often enough for the savings to accumulate. Brynjolfsson has since described this shift more broadly: the economy is transitioning from an era of AI experimentation into one of structural utility, where the gains are no longer theoretical but visible in real workflows and real performance data. That transition does not require a technical background. It requires the willingness to attach a tool to a real task and see what happens.

There is one more detail from Brynjolfsson's research worth highlighting, as it speaks directly to the question of who benefits from AI. The study found that the AI tool essentially captured the patterns of the best-performing agents and made those patterns available to everyone.

The least experienced workers improved because they gained access to a version of institutional knowledge that would normally take months or years to absorb. In a business context, this means that AI tools are not just productivity boosters. They are equalizers. They reduce the gap between the newest team member and the most seasoned one, not by replacing expertise but by accelerating the path to practical fluency. For someone entering a new field or building new capabilities mid-career, that compression of the learning curve is one of the most valuable things AI can offer.

The same pattern holds across other workflows. Meeting notes are easier to process when the first pass is already generated, and your job shifts to reviewing and correcting rather than starting from scratch. Research across multiple sources moves faster when you begin with a synthesized view and go deeper only where it matters. Presentation structure comes together more quickly when the outline is already in front of you, and you can spend your time on the story and the decision you are trying to drive. In each case, the workflow becomes repeatable, which is a more important outcome than any single time saving.

This is another idea that has appeared before in this book, and again, it is repeated because it works. When I want to know whether I truly understand a workflow, I try to explain it to someone else. If I cannot walk through the steps clearly, the workflow is not stable yet. Teaching exposes that faster than any amount of private practice.

Choosing Tools Without Getting Lost

Once you accept you cannot keep up with everything, the question changes. The goal is not to track every release. The goal is to choose tools that reduce the effort required for repetitive work.

The question I ask before trying anything new is always the same: Does this solve something that is getting in my way this week? If the answer is no, I close the tab. Beyond that, I check whether I can get value without hours of setup, whether the tool fits into what I already use, whether it is safe for the type of data involved, and whether it looks stable enough to still exist in six months. If a tool clears those filters, I give it a short trial with a measurable target. Saving time is the obvious one. Reducing rework is another. Improving the quality of first drafts is another. If the trial produces a clear improvement, I write down the workflow in my notes. Not a long document. Just enough so I can repeat it later without having to relearn it from scratch. There is a reason I am strict about documenting what works. In the early AI phase, the biggest risk is not missing a tool. The bigger risk is that you keep starting over because you never turn the learning into something repeatable.

My toolkit has changed several times while writing this book. Tools I relied on six months ago have been replaced. Others that felt optional became central once I found the right workflow. Listing a specific stack here would suggest there is a single correct set of tools to adopt. There is not. What matters is the process of choosing, not the result.

I will say this much about my experience. I use different tools for different types of work. One handle writing and analysis well. Another is better suited for long documents and complex reasoning. A third works best within the applications I already use at work. A fourth excels at research and fact-checking. I also use tools for visual design, voice, video, and orchestrating multi-step projects. Each earned its place through trial, not hype. Several others did not survive the test. The reason I am keeping this general is simple.

AI tools evolve quickly, and a printed list becomes outdated within months. Instead of freezing a snapshot here, I maintain a current version of my working toolkit at adapterslab.ai/tools, where it is updated as things change. That page includes not just the tools but the specific workflows I use them for, so you can see how each one connects to a real recurring task rather than a feature list.

The principle is more durable than any tool list: tie a tool to a recurring task, run it long enough to get a stable result, and drop it if it does not hold up. If you apply that filter to your own stack, the specific names matter less than the discipline of testing and keeping only what works.

In Part III of this book, you will find a closer look at the major AI platforms and how they are changing careers and workflows across industries. That chapter goes deeper into what each system does well, where it falls short, and what it means for the work you do every day. If you want the detailed view, that is where to look. For now, the point is simpler: your stack should serve your work, not the other way around.

Making It Repeatable

Interest is rarely the problem because a tool gets tested once, looks promising, and then the demands of real work take over before the habit has had time to form, which is why the rhythm that works for me is simple. Try the tool on real work in the first week.

In the second week, tighten the workflow and document the steps so I can repeat them without having to reconstruct them each time. In the third week, teach it once, even informally, because explaining something to someone else forces a clarity that practicing alone rarely produces. In the fourth week, decide whether it stays because it consistently makes the work easier, or whether it leaves before it becomes clutter. A more detailed version of this structure, with prompts and decision criteria, is available at adapterslab.ai/tools.

This mirrors the argument in AI and the Future of Work: adoption occurs when humans and AI collaborate in small, repeatable ways, not when organizations try to roll out every new tool at once.

The science of habit formation supports this. BJ Fogg, a Stanford University behavior scientist and founder of the Behavior Design Lab, spent more than twenty years studying why people succeed or fail at building new behaviors. In Tiny Habits, he proposed a model that explains behavior change with a simple formula: behavior happens when motivation, ability, and a prompt come together at the same moment. If any one of those three elements is missing, the behavior does not happen, no matter how strong the others are.

Fogg's core insight, and the one that matters here, is that people consistently overestimate the role of motivation and underestimate the role of ability. They assume that if they care enough, they will follow through. In practice, motivation is unreliable. It rises and falls with energy, mood, and competing priorities. What drives lasting behavior change is making the new action so small and so easy that it barely requires motivation at all. This is why the four-week rhythm I described works as well as it does. In the first week, the action is small: use the tool on one real task. That is a tiny behavior, anchored to the work you are already doing. The prompt is the task itself, the report that needs writing, the meeting notes that need summarizing, and the data that needs organizing.

The ability threshold is low because you are not trying to master the tool. You are just trying to get one result. Fogg calls this "making it so easy to do that it is hard to avoid." He also studied why products like Instagram and Google succeeded at scale and found the same principle at work: they helped people do what they already wanted to do and made it extremely easy to start. The same logic applies to AI tools. If the first use requires hours of setup, most people will not come back. If the first use produces a useful result in under 10 minutes, they will almost certainly.

Fogg also emphasized something that runs counter to how organizations approach training: people change best when they feel good, not when they feel bad. Shame, pressure, and fear of falling behind are poor foundations for sustained behavior. They might create urgency, but they do not create habits. What creates habits is the experience of success, especially early success. When you use an AI tool on a single task and see a clear improvement, that small win generates a feeling of capability. That feeling makes you more likely to try the tool again, and the repetition is what turns a trial into a workflow.

This is why I suggest teaching the workflow to someone else in the third week. Teaching is not just a test of understanding. It is a celebration of progress. It forces you to articulate what you learned, which reinforces the habit, and it gives you feedback that tightens the process. Fogg would recognize this as a form of what he calls "wiring the habit into your brain" through positive reinforcement rather than obligation.

Fogg's work also offers a useful warning about what does not work. Large-scale behavior change programs that rely solely on motivation, the kind organizations often roll out with fanfare and mandatory training sessions, tend to produce short bursts of activity followed by rapid decline. The reason is structural, not motivational. When the new behavior is difficult to do, and the prompt is disconnected from existing routines, even motivated people revert to their defaults within weeks. Fogg's solution is to anchor new behaviors to existing ones.

In the context of AI adoption, this means the best time to use an AI tool is not during a dedicated "learning session." It is during the task you were already going to do. When you pair AI with the moment you would normally open a blank document, or start a research pass, or draft a set of meeting notes, the tool becomes part of your existing flow rather than an addition to your schedule. That is how a trial becomes a habit, and how tech empowerment moves from intention to practice.

The most common reasons people stop making progress are worth naming briefly, because recognizing them makes it easier to move past them. Perfectionism slows adoption by creating a need to understand everything before starting. With AI tools, the approach rarely works: the software changes quickly, and the target understanding keeps moving. Waiting for formal training creates a similar delay.

The belief that using AI tools requires technical ability conflates two different things: knowing how a tool works internally and knowing how to use it well in a specific context. Fluency matters more than certification here. If you can run a workflow reliably and explain it clearly, you are already ahead of most people in your organization. And adding too many tools at once, something I learned the hard way when I tried to integrate several AI note-taking tools simultaneously, creates more to manage rather than less. One solid workflow would have been enough.

This is why Tech Empowerment sits at the end of ADAPT. Awareness helps you see change early. Direction helps you choose where to aim. Action creates evidence. Positioning helps that evidence travel. Tech Empowerment is what makes the new way of working easier to repeat, so progress does not depend on motivation or novelty.

Pick one recurring task. Attach one tool to one slice. Measure the difference. Write down what worked. If it holds up for a few weeks, share it with one person. That is enough to turn curiosity into capability.

PART III

WORKING THE FUTURE

Application and scale. Readers learn how to embed AI into daily workflows, build personal systems, and navigate corporate realities, including guardrails and adoption gaps. It closes with use cases, tool guides, and a forward view on careers, including how to keep learning, ship prototypes, and evolve with the next wave.

10

THE ADAPTER'S TOOLKIT: A PROFESSIONAL LIFE IN PRACTICE

Before we examine the platforms and the changes reshaping work across industries, I want to show you what an adapted professional life looks like, not in theory, not on a keynote slide, but in the lived, imperfect, occasionally frustrating reality of someone doing the work every day.

Elena is a marketing operations manager at a consumer goods company with about two thousand employees. She has been in the role for three years and in the industry for fifteen. She does not have a computer science degree, a background in data science, or any affinity for technology as a discipline. What she does have is a working system that makes her better at her job than she was eighteen months ago. She built it using the same method this book has been describing all along: one small experiment at a time, with honest assessment of what worked and a willingness to abandon what did not. Her story is not remarkable. That is why I am telling it.

Each of the following sections traces one of the five ADAPT principles through the details of Elena's actual practice. The principles do not operate in a vacuum in her work, and they will not in yours. But tracing each one separately makes visible what the integrated version can otherwise obscure: that behind every efficient workflow is a series of deliberate choices, many of them unglamorous, about what to invest in, what to question, and what to protect from optimization altogether.

Awareness: Knowing Before it Becomes Obvious

One of the most consistent findings in organizational research is that professionals who operate with strong situational awareness, a clear sense of what is happening at the edges of their industry, not just at the center, tend to make faster and better decisions when it matters. The reason is not that they are smarter. It is that they are less surprised. They have already integrated new information into their thinking before it becomes urgent.

Elena understood this intuitively long before she had a system to act on it. What she lacked was a way to build and maintain that awareness without it consuming disproportionate time. The old approach, manually scanning industry publications, monitoring competitor activity, and checking regulatory updates, was not unreliable so much as it was unsustainable. It required blocks of focused time that her role did not offer. When those blocks were crowded out, as they usually were, her situational awareness degraded.

The turning point came when she stopped trying to carve out time for awareness and started building it into existing habits. She set up a custom digest using Perplexity AI that tracks three things: her company's brand mentions, competitor product launches, and regulatory changes in her sector. Reading it takes roughly ninety seconds. She does it while making coffee, before opening her laptop. She does not act on most of what she reads. The point is not to respond; it is to know when something important eventually requires a decision. She has already absorbed the relevant context rather than encountering it under pressure.

She tried simpler alternatives first. RSS feeds required too much manual curation to stay relevant. Curated newsletter services were better, but always slightly misaligned with her specific concerns.

What made Perplexity work for this purpose was the ability to define queries in plain language and refine them over time, so that the digest reflected her actual professional terrain rather than a generic approximation of her industry.

The tool earned its place because it reduced the gap between the effort she put in and the signal she got out. Every tool she uses gets evaluated against that same standard. Once a week, usually during a meal break, she runs through three rotating questions covering emerging trends in her category, a new tool or capability she has come across recently, and changes in consumer behavior. She calls it a signal scan, and it takes less time than the average meeting.

Most weeks, the answers confirm what she already suspects. Occasionally, something surprises her. When she reads about a European company using AI-generated, personalized packaging tied to regional purchasing data, she does not immediately know whether it is relevant to her company's roadmap. She saves it anyway; in a Notion knowledge base, she has tagged it by theme. The knowledge accumulates. Over fourteen months, this habit has compounded into a kind of contextual richness that is difficult to attribute to any single search, yet whose absence would be immediately felt.

Her colleagues have started asking where she finds things. The answer is fifteen minutes a day, consistently. Knowledge does not arrive all at once. It builds the way trust does, through repeated small deposits rather than a single large transfer.

Direction: Knowing What Your Judgment Is Actually For

Direction, as discussed earlier, is not only about career trajectory. In daily practice, it means deciding which decisions genuinely require your thinking and which ones can be handed off without losing anything important. Most people who start using AI tools skip this question entirely. They automate what is easy to automate and move on.

Elena learned this early in her experimentation. Like most people who get excited about AI tools, she moved quickly: testing, adopting, and automating across several workflows at once. One of them was her team's monthly performance updates. The AI drafts were efficient, structurally sound, and completely unremarkable. She used them for three months before realizing what she had given away without meaning to. Writing those updates had been how she formed her own views about how her team was doing. Once a draft appeared on screen, her job became editing rather than thinking. The thinking, she understood too late, had been the point.

She pulled back. She now writes those updates herself, using AI only to check for gaps in her reasoning once she has worked through it. It takes longer. It is worth it.

The same issue appeared in other experiments. Automated scheduling created more coordination overhead than it resolved. AI-generated social copy for the company's LinkedIn was technically acceptable but consistently forgettable: it read like a press release written by someone who had studied many press releases without understanding why they existed. A meeting note synthesis tool was useful around 60% of the time and was confidently wrong the other 40%. The errors were harder to catch precisely because the summaries sounded authoritative, and when something sounds authoritative, the natural impulse is to trust it. That impulse is where the risk lives.

None of these pullbacks was permanent. She revisited the meeting notes tool six months later, found it had improved substantially, and reintroduced it with a verification step built into her process. The principle she now operates by is simple: a tool earns its place by making things genuinely better, not just faster. Speed that removes the thinking that sharpens your judgment is not an advantage worth taking.

This is what separates professionals who use AI well from those who simply use it a lot. The difference is not how much they automate. It is how clearly they understand what they are choosing to keep.

Action: From Idea to Evidence Without Waiting

There is a persistent friction in organizational life between having an idea and being able to test it. The conventional cycle that involves brainstorming, gathering requirements, writing a proposal, waiting for approval, assigning to development, and iterating can take months to produce something a room full of people can look at and respond to. By then, the conditions that made the idea worth testing may have changed. The energy that made it feel urgent has usually dissipated. Research on organizational decision-making consistently finds that the speed of testing an idea is as important as the quality of the initial idea, because tangible prototypes surface real problems that abstract proposals cannot anticipate.

Elena had experienced this cycle enough times to feel genuine frustration with it. When her team identified a need for a new customer feedback workflow, the default would have been to schedule a brainstorming session, commission a requirements document, and wait six to eight weeks before anyone could see what the thing might look like. She proposed a different approach: 90 minutes with a colleague to build a working prototype using Lovable.

By the end of that session, they had a web form that collected customer feedback, routed responses by sentiment category, and generated daily summaries for the relevant team leads. It was not polished. The sentiment routing was clumsy with ambiguous responses. Two of the summary categories overlapped in ways that would confuse anyone trying to use them. But those were solvable problems, and more importantly, they were now visible problems. You cannot fix a flaw in a proposal. You can fix a flaw in a prototype, because the prototype shows you what the proposal was only describing.

She chose Lovable specifically because it allows non-technical professionals to build functional interfaces without writing code, and because the output is something she can hand to a developer as a basis for refinement rather than a description of what they should build from scratch. The choice compressed the distance between imagination and feedback. That compression is what Action looks like in practice: not perfection on the first attempt, but speed to the point where real learning becomes possible.

There is a broader principle here that extends beyond any single project. The willingness to show something imperfect to move from concept to demonstration before you are ready requires a specific kind of professional confidence. Not the confidence that comes from certainty, but the kind that comes from repeated evidence that imperfect starting points lead to better outcomes than polished proposals that never get tested. Elena built that confidence incrementally, through a series of small experiments that went well enough to justify the next one. Trust in a process, like trust in a tool, is not installed all at once. It accumulates through use.

Positioning: The Human Layer That Makes the System Work

Of all the things AI tools can do well, managing professional relationships is not one of them. They can generate competent text. They cannot know the history behind a vendor relationship, how a colleague handles difficult feedback, or the organizational politics that make a straightforward message land very differently depending on who sends it. Positioning, in this context, means knowing where your judgment has no substitute and protecting that space rather than gradually giving it away to convenience.

Elena discovered this principle through a minor failure that was instructive in proportion. Early in her use of Copilot for email drafting, she sent a reply to a long-standing agency partner with only light editing. The message was professionally worded and factually accurate.

The agency lead called her that afternoon. The tone had felt cold, he said. Transactional. "Not like you." She spent twenty minutes on a follow-up call repairing what five minutes of careful editing would have prevented. She has not repeated the mistake, and the lesson she drew from it goes beyond email: the tools in her stack reflect the inputs she gives them. They do not know her, her relationships, or the context those relationships carry. That knowledge is hers to apply, and applying it is not optional.

She encounters the same issue at a larger scale whenever she uses AI-generated material in high-stakes internal communications. When she used her ChatGPT Project to draft talking points for an all-hands presentation on the company's brand refresh initiative, the output was accurate, well-structured, and almost entirely wrong for the occasion. The problem was not factual. The problem was the register. The talking points read like a strategy memo, which was exactly what she had been feeding the model all morning. They answered the question "what is changing?" when the room of three hundred employees would be asking "why should I care?" and, more pressingly, "is my job safe?"

She scrapped the draft and opened a blank document. She wrote the opening remarks herself, then returned to the model with a different prompt: not "draft talking points for this presentation" but "given this audience and this anxiety, what are the three questions most likely to be on people's minds"? That reframing produced something genuinely useful. She rebuilt the presentation around those questions.

The insight she extracted from this failure, and has since internalized as a working principle, is that *AI reflects the frame you give it*. If the frame is wrong, the output is wrong regardless of how polished it sounds. Recognizing a bad frame requires the kind of contextual judgment that only comes from knowing the people involved, the relationship, and the stakes involved. That is not a skill that any tool can replicate. It is precisely what fifteen years of professional experience looks like in practice.

Tech Empowerment: Building a Toolkit That Amplifies What You Already Know

The final principle is the one most people reach for first and the one that works least well when approached in isolation. The goal of Tech Empowerment is simple: fewer tools, used well, applied to the problems where they make a difference, with a clear understanding of what each one is designed to do and where its usefulness ends.

Elena's current set of tools is small. Perplexity for information synthesis, Copilot integrated into her existing email workflow, a ChatGPT Project built specifically around her campaign analysis needs, Lovable for rapid prototyping, and Claude as a reflective journal. Five tools, each chosen for a specific purpose, each selected after a trial period in which she tested it against a real problem rather than an imagined one. She has tried and discarded twice as many tools as she currently uses. That culling is not a failure. It is judgment.

The centerpiece of her setup is the ChatGPT Project she has built for campaign analysis. It is loaded with her team's performance data, brand guidelines, historical benchmarks, and insights into her VP's decision-making priorities. When she needs to prepare for a performance review meeting, she uses it as a thinking partner: asking follow-up questions, testing alternative framings, surfacing counterarguments she had not considered.

The output would take ninety minutes to build from scratch. She spends twenty minutes reviewing it, adding the kind of contextual observation that no dataset captures, and shaping it into something that reflects her perspective rather than merely aggregating information she already had access to.

She chose ChatGPT Projects over a shared analytics dashboard precisely because the Projects interface supports a kind of interactive reasoning that dashboards do not. She is not reading someone else's summary of the data. She is thinking out loud, with a tool that can keep up. The distinction matters: tools that let you engage with information actively tend to sharpen judgment, while tools that simply present information can, over time, create a kind of passive dependency on other people's interpretive frameworks.

The reflective journal is perhaps the least technically obvious of the tools she uses, and in some ways the most important. Each evening, before closing her laptop, she spends ten minutes writing in a Claude conversation the way some people write in a notebook: what worked, what did not, and one question she wants to carry into tomorrow. The entry is not long. It does not need to be. The act of articulating what went wrong, precisely and honestly, is itself the cognitive work.

Cognitive science research on reflective practice consistently finds that experience without structured reflection produces far less learning than the same experience with it. What Elena has built into the end of her working day is not administrative overhead. Without that step, experience accumulates without turning into anything more useful than the next experience.

It took about three months for the practice to stop feeling like an added obligation.

She used a phone reminder for the first six weeks to keep it from slipping. Now she does not need one. On the days she skips it, she finds herself returning to the same questions later, which tells her more about its value than any deliberate test would.

What the Whole System Actually Is

It is worth pausing here to name what Elena has built, because it is easy to mistake the individual components for the thing itself.

The thing itself is not a set of tools. It is a practice of deliberate experimentation combined with sound evaluation: try something specific, assess whether it genuinely improves the relevant work, keep it if it does, discard it if it does not, and revisit discarded options when circumstances change. Each of the five ADAPT principles operates within this practice rather than alongside it. Awareness guides what she looks for.

Direction governs what she protects from automation. Action shortens the gap between idea and evidence. Positioning preserves the relational intelligence that tools cannot replicate. Tech Empowerment ensures that the tools she chooses amplify her existing strengths rather than substituting for judgment she has not yet developed. No single tool transformed her role. No single week was the turning point.

Over several months, a series of small decisions accumulated, each modest on its own, but together they redirected how she spent her cognitive energy.

The gain was not in automating more tasks but in developing a clearer sense of where her own judgment adds something that no tool can replicate.

That clarity is available to any professional willing to build toward it, as Elena did: one honest experiment at a time.

A Bridge to Your Own Practice

Before you move to the next chapter, take five minutes with a single question: *What is one recurring task in your professional week, something you do every time a certain kind of request arrives, every time a particular meeting approaches, every time a specific pressure builds, that resembles something in Elena's practice?*

It does not need to be an exact match. It might be a version of her morning awareness scan, or a preparation ritual before a recurring meeting, or a communication task you handle manually because you have not yet explored whether there is a better approach. It might be a reporting task you dread, or a synthesis process that takes longer than it should, or a type of decision you make slowly because you are working from incomplete information.

Choose one. Then design the smallest possible experiment you could run this week. Not a new system. Not a workflow overhaul. One task, one tool, one week, and one simple assessment afterward to see whether the work improved. Not whether the tool was impressive, not whether it saved time in the abstract, but whether the actual output was better and whether the experience of producing it felt like the kind of professional you want to be.

That is how Elena started. Not with a plan, not with a training program, but with one experiment on one real task. The difference between where she is now and where most professionals remain is not talent or technical ability. It is simply the number of experiments she was willing to run. Run the first one.

11

WHEN AI GETS IT WRONG

This chapter exists because working seriously with AI also means understanding where it falls short. If the rest of this book is about what AI makes possible, this one is about its limits, and what those limits mean for the professionals who depend on it. The goal is not to discourage enthusiasm but to build the kind of judgment that makes the tools more useful: knowing when to trust them, when to push back, and when to set them aside and think for yourself.

The Confidence Problem

The first thing anyone who works with AI should learn is that these systems sound confident. Whether the output is correct or completely wrong, it arrives in the same calm, polished tone. There is no hesitation, no qualifier, no visible sign that the answer you just received is half-invented.

In 2023, two attorneys in New York submitted a legal brief containing six fabricated case citations, cases that did not exist, with plausible-sounding names and docket numbers. They used ChatGPT for legal research and did not verify the output. The judge sanctioned both lawyers.

In my own work, I have seen AI produce risk assessments citing reports that were never published, document summaries that missed critical clauses, and financial projections built on unrealistic assumptions. In every case, the output looked professional. The errors were in the facts.

What makes this especially hard to manage is that confidence and accuracy do not move together. Humans are wired to trust fluent, articulate communication. That instinct served us well for most of history, because fluency was often a reasonable sign of knowledge. A person who sounded certain usually had some basis for that certainty. AI breaks that assumption completely. The output comes across as polished and authoritative, whether the underlying claim is solid or completely made up. There is no tonal difference between the two.

The lesson is simple: AI output is a draft, never a deliverable. The moment you treat it as finished work, you have handed over your judgment to a system that has none.

Pattern Recognition for Practice: The confidence of AI output is a design feature, not a quality signal. Train yourself to read AI-generated work the way a good editor reads a first draft: with appreciation for what is there and careful suspicion about what might be wrong.

Hallucination: The AI Error That Looks Like a Finished Answer

The technical term for AI producing false information is *hallucination*. The system generates text that fits language patterns, but there's no built-in way to check whether it matches reality.

Researchers at Harvard Kennedy School have shown that hallucinations are not glitches but a structural feature of how large language models work. The model is not searching a database and retrieving a fact. It predicts what text is likely to come next based on patterns in its training data. When the training data contains enough relevant signal, the prediction is often correct. When it does not, the model fills the gap with something that looks right but is not. It has no internal mechanism that registers when it has crossed that line.

This matters in practice because hallucinations are most common exactly where training data is thin or changing quickly. Ask an AI for the founding date of a well-known company, and it will probably be correct. Ask about a niche regulatory change in a small market from last quarter, and you may get something completely fabricated but indistinguishable from the truth. The higher the stakes and the more specialized the domain, the higher the risk. That is not a coincidence. It is built into how these systems work.

For professionals, this creates a specific problem: the areas where you most need AI's help are often the areas where it is most likely to get things wrong. Niche regulations, recent statistics, specific citations, numerical calculations, and anything that sits at the intersection of several specialized fields all carry a higher risk of hallucination. Knowing this does not mean avoiding those uses. It means checking them more carefully.

The discipline this requires is straightforward: use AI to generate first drafts and initial analyses, then review the output as you would a talented but inexperienced junior colleague. Appreciation for the speed and close attention to every factual claim.

It also helps to build a personal sense of where the risk is highest. Not every output carries equal weight. The ones that do deserve verification before they become the basis for a decision.

Pattern Recognition for Practice: Build a personal list of high-risk task categories where AI hallucination is most likely. Treat output in those categories with an extra layer of review, not because the tool is useless there, but because the cost of a missed error is highest.

Bias: The Patterns We Did Not Choose

AI systems learn from data, and data reflects the world as it has been, not as it should be. A 2025 IBM-backed survey found that more than forty percent of companies using AI recruiting tools detected bias in shortlists, even after debiasing efforts. Stanford researchers demonstrated that leading resume-screening models systematically favored certain demographic profiles.

I have seen AI evaluations that consistently rated organizations or candidates from certain regions lower, not because of objective performance data but because the training material contained more negative coverage about those regions. The model learned an association and reproduced it without realizing the pattern stemmed from uneven reporting rather than actual quality. No one in the workflow flagged it because the output appeared to be a normal score.

This is what makes bias harder to catch than hallucination. A fabricated citation is wrong in a way that can be verified. A biased score is wrong in a way that requires asking what a fair score would look like, and most people reviewing AI output do not ask that question.

Bias also appears in outputs that professionals do not examine closely enough. AI-generated summaries, stakeholder communications, and market analyses can carry cultural and linguistic assumptions from the training data without it being obvious. The writing can sound neutral while reflecting perspectives that are anything but.

The answer is not to avoid AI. It is to approach it with respect for what it can do and vigilance about what it carries. Diverse review panels, explicit prompting for alternative perspectives, and routine audits of AI-assisted decisions in high-stakes areas are not bureaucratic excess, but the minimum standard of responsible use.

Pattern Recognition for Practice: When using AI for evaluation, screening, or analysis involving people or places, build in at least one step where you explicitly ask the tool to identify assumptions it may have made. A simple prompt addition, such as 'what perspectives or groups might this analysis underrepresent?' will not catch everything, but it makes the question part of the process rather than an afterthought. The outputs that look most neutral are often the ones most worth examining.

What You Lose When AI Does the Thinking
When Delegation Becomes Decline

There is a less visible risk that causes a gradual decline rather than a visible breakdown. When you hand routine thinking to AI, you gain speed but lose the practice that keeps your judgment sharp. The professional who stops reviewing important documents closely may lose the ability to catch the clause AI missed. The analyst who stops building models from scratch may lose the instinct that tells them when an output does not make sense.

An MIT Media Lab study, reported in the Harvard Gazette, warns that heavy reliance on AI assistants can lead to cognitive atrophy, much like GPS has weakened everyday navigation skills by taking over what used to be routine mental work. Most people who use GPS regularly have noticed this: routes they drove for years now require the app because the mental map never got maintained. The same pattern can emerge in professional thinking, and it tends to happen without any single moment you can point to.

The research goes further, noting that both lower-order skills, such as memory and recall, and higher-order capacities, such as critical thinking and strategic reasoning, are at risk when AI handles the mental work that used to exercise them. Speed and capability are not the same thing. You can become faster at producing output while also becoming less able to judge it.

A 2025 Harvard Business Review article found that organizations with the highest levels of AI delegation saw declines in strategic thinking quality, not because the tools were poor, but because people stopped exercising the judgment that produces original insight. Output volume increased. Thinking quality did not keep pace.

The ADAPT method addresses this through Tech Empowerment. The goal is not to use AI for everything. It is to use AI where speed matters more than depth, and to protect the parts where your judgment is irreplaceable. That requires an honest, ongoing review of which thinking skills you are still using and which you have handed off without a conscious decision.

Most professionals who have been using AI daily for a year or more have not yet done that review.

Pattern Recognition for Practice: Identify two or three skills that are genuinely irreplaceable in your role, the judgment calls that make you valuable rather than just productive. Continue using those skills even when AI could technically handle the task. The point of delegation is to free up your best thinking, not to replace it.

The Prompt Engineering Illusion

There is a gap in how many professionals think about AI risk, and it is worth naming directly. Learning to write effective prompts is a real skill. It produces better, more relevant, more structured output. It is also not the same as managing AI risk.

Professionals who become skilled at prompting sometimes develop a confidence that they have learned to control the tool. They have learned to guide it. That is different. A well-written prompt improves the fluency and relevance of the output. It does not reduce hallucination risk in the underlying model. It does not correct structural bias in the training data. It does not make a fabricated citation real.

The danger is that prompt skill creates a false sense of security. You feel more in control because the output more closely matches what you asked for. But what you asked for and what is true are separate questions, and no amount of prompting closes that gap. The verification discipline described later in this chapter applies just as much to a polished, well-prompted output as it does to a rough first attempt.

Pattern Recognition for Practice: The next time an AI output closely matches what you asked for, pause before accepting it. A response that fits your prompt well is easier to trust and harder to scrutinize. That is precisely when verification matters most. Check the claims that sound most authoritative, not the ones that seem uncertain. Those are the ones most likely to pass through unexamined.

When to Stop Using AI Entirely

Most conversations about AI risk end with some version of 'use it carefully.' That is true, but it assumes careful use is always possible. For some kinds of work, it is not.

Consider high-stakes legal interpretation, where the margin for error is narrow, and the consequences of a missed nuance are severe. Or sensitive HR decisions, where the details of an individual case require contextual human judgment that no AI system can reliably replicate. Or original ethical reasoning, where the point is to think through a problem from first principles, and where an AI draft shapes your thinking before you have had the chance to form your own view.

The anchoring effect is underrated here. When you read an AI-generated first draft before forming your own opinion, you are more likely to end up near that draft's conclusions than you otherwise would. In fields where independent judgment is the core value of your work, anchoring is a problem you cannot resolve after the fact, because you would need to know what you would have concluded without the draft in front of you. You cannot unknow what you have already read. This does not apply to most professional tasks. But every professional has a few categories of work where the quality of their independent thinking is what they are being paid for. Those categories deserve a deliberate rule: no AI-first drafts. Start with your own thinking. Use AI later, if at all.

Pattern Recognition for Practice: Identify the work in your role that requires truly independent judgment, the decisions or analyses where your unshaped thinking is the product. Protect those tasks from AI involvement at the drafting stage.

A Plausible Lie

In one case, AI was used to check a claim about a company's expansion into a new market. The tool produced a detailed response that covered the timeline, investment figures, and the strategic rationale behind the move. It was exactly the kind of output that gets forwarded without a second thought. When the information was checked against external sources, it was found that the expansion had never occurred. The AI had constructed a plausible narrative from fragments of real events.

How close it came to being missed is worth acknowledging directly. The output was well-structured, the figures were plausible, and the strategic rationale sounded exactly like something a real company would say. Nothing in the text signaled a problem. The only reason it was caught was that it was verified independently. The AI offered no warning, no hesitation, no indication that anything was wrong.

That is the real lesson. Uncertainty will not always announce itself when AI gets something wrong. The output will often feel confident and complete. The verification practice described below is not for situations where something looks suspicious. It is for the moments when everything looks fine, because that is when a wrong claim makes it into a report, a presentation, or a decision that cannot easily be walked back.

Building a Verification Practice

Develop a personal verification protocol and use it every time an AI output leaves your desk.

For factual claims, check the source. Not the AI's summary of the source, the source itself. This takes longer than trusting the output, and it is not optional for anything consequential.

For citations and references, search for the actual document. Confirm the author, publication, date, and that the quoted or paraphrased material appears in the work cited. This step alone would have prevented the attorney sanctions described earlier in this chapter.

For analytical output, apply the common-sense test: does this conclusion make sense based on what you already know? Where the AI's analysis conflicts with your prior understanding, that conflict is a signal to investigate, not a reason to change your view automatically. For numerical output, run spot checks. AI systems can handle arithmetic, but they also confidently produce numbers that are off by an order of magnitude, or that rest on assumptions buried so deep in the prompt that no one notices them.

For creative or strategic work, ask the harder question: Is this genuinely original thinking, or is it a polished version of the most common answer to this kind of prompt? If everyone using the same tool with the same prompt gets roughly the same output, that output is not a competitive advantage. Your job is to push past it.

None of this means AI is not worth using. It means the value of using it well comes partly from knowing exactly where to be skeptical. The tools handle the volume. You handle the truth.

12

THE TOOLS THAT ARE CHANGING HOW WE WORK

The conversation about artificial intelligence at work almost always begins with ChatGPT. That makes sense. For many professionals, generative AI was the first moment when the technology stopped feeling abstract and started feeling personal. A marketer could ask for a campaign outline. A lawyer could summarize pages of legal text. A manager could draft a memo in seconds.

But the story unfolding inside organizations is much larger than one chatbot. AI is gradually embedding itself across the entire workday, from how we write and research to how we build tools, manage schedules, and make decisions. Each of these capabilities may appear modest on its own. Taken together, they begin to reshape how work flows through an organization and, more importantly, how careers evolve within it.

This chapter is organized into two parts. The first looks at the major AI platforms through a practical lens: what each one is good at and what it means for how you work and how your career develops. The second maps the broader landscape of AI systems being embedded in the workplace, from automation and scheduling to research, design, and personal knowledge management. Together, they offer a working picture of where AI is showing up today, not in theory, but in the daily routines of professionals across industries.

For more detailed and continuously updated profiles of these platforms, including feature comparisons, limitations, and practical workflows, visit adapterslab.ai/vanguard.

Part One: The Platforms That Matter for Your Career

ChatGPT: The Partner You Did Not Know You Needed

Most people's first experience with ChatGPT is small. A rewritten email. A summary of meeting notes. A brainstorm that gets unstuck. It feels like a shortcut, and for a while, that is all it is.

The real change begins when you use it in your actual work, not in demos or side experiments, but in the tasks that fill your week. You paste a contract into a tool and ask it to identify the risk clauses. You give it notes from three meetings and ask for one clear action list. You describe a problem you have been trying to solve for days and ask for five possible approaches.

The result is not always perfect, but it gives you a useful starting point that would have taken much longer to create on your own. Over time, that changes how you show up. You walk into meetings with cleaner analysis. Your emails are sharper. Your proposals are more structured. You are not doing less work. You are doing different work, the kind that requires judgment, context, and the ability to make a case rather than just compile information. That is the real value of ChatGPT in a career. It compresses the time spent on production so you can spend more time on the thinking that moves things forward.

For any professional stuck in a cycle of formatting reports, chasing approvals, and managing administrative overhead, ChatGPT is the tool that begins to free up the space for strategic contribution. Not because it replaces your expertise, but because it handles the parts of your day that never required your expertise in the first place.

Claude: The Thinking Partner for Complex Work

Claude enters your workflow differently. Depending on the model and version you use, where ChatGPT is fast and versatile, Claude is patient and thorough. It reads long documents, maintains context across many pages, and delivers the kind of structured reasoning that is hard to get from a quick prompt.

If you work with large files, complex strategies, category reviews, policy documents, research syntheses, or anything that requires holding a lot of information in your head at once, Claude is where the value currently shows up. You paste in your supplier scorecards, market trends, and internal demand forecasts, and it surfaces patterns you might not have noticed: a recurring risk in a specific region, a cost-saving opportunity buried in the data, or a supplier that is underperforming but has not triggered alarm thresholds. You walk into the review better prepared, not just with data, but with a narrative that links the pieces together in a way that makes sense to people outside your function.

That kind of preparation changes how you are perceived. You begin to be seen less as someone who manages a category and more as someone who understands the business. Claude does not replace the relationships you have with stakeholders. It gives you better material to bring into those conversations. Over time, that pulls your role from execution toward influence, which is exactly where careers grow in an AI era.

Gemini: The Connector Across Your Systems

Gemini works best when your daily work already lives within an ecosystem of connected tools, especially Google Workspace. It does more than answer questions. It contextualizes them within the flow of your applications, from documents and spreadsheets to email and calendar.

For professionals who spend much of their day moving between systems, pulling data from one place and summarizing it in another, Gemini reduces that burden in corporate settings. You ask it to summarize last quarter's activity by category and region, and it returns a concise narrative you can use to start a conversation with leadership. You no longer need to wait for a

BI team to build a dashboard or for an analyst to run a custom report. You can do the first pass yourself.

That self-service capability is powerful. It makes you more agile because you can answer your own questions faster. And it makes you more influential because you bring insight to conversations instead of waiting for someone else to provide it. In a career context, Gemini is the tool that helps you stop working in isolation within your function and start connecting your work to the broader business, including demand planning, finance, risk management, and strategy. That cross-functional view is increasingly what separates people who manage tasks from those who shape decisions.

Copilot: The Layer Inside Your Workday

Copilot does not require you to change how you work. It sits inside the tools you already use: Word, Outlook, Excel, Teams, and PowerPoint. It summarizes meeting transcripts, drafts email replies, interprets data in spreadsheets, and generates slide outlines from rough notes.

Adoption data from Microsoft suggests that 86% of Copilot users have incorporated it into their daily tasks. That number matters less as a headline and more as a signal: when AI is embedded in the applications people already open every morning, adoption happens without a deliberate decision to "use AI." It simply becomes part of how work gets done.

For your career, this matters in a specific way. Copilot makes you faster and more consistent without requiring you to learn a new platform or change your habits. You respond to stakeholders more quickly. Your documents are cleaner on the first pass. Your meeting follow-ups go out the same day rather than 3 days later. None of that sounds dramatic, but it compounds. The person who consistently delivers clear, timely, well-organized work builds trust faster. And trust is what opens the door to bigger projects, broader scope, and the kind of visibility that moves careers forward.

Lovable: The Builder's Shortcut

Lovable is different from the other tools in this chapter. It does not help you write, research, or analyze. It helps you build. You describe what you want, such as an app, a dashboard, a workflow tool, or a prototype, and it generates a working version using real code that you can export, modify, and own.

I used Lovable to build adapterslab.ai, the platform that accompanies this book. I had no coding background. What I had was a clear picture of what I wanted and the willingness to iterate. I would describe the feature in plain language, review the tool's output, and refine the prompt until the flow made sense.

For anyone who has ever had an idea for a tool that could solve a real problem at work but lacked the budget or the technical team to build it, Lovable changes the math. You can go from concept to prototype in days rather than months. That matters for career growth because it puts you in a position most professionals never reach; you are no longer just identifying problems; you are building solutions. In organizations that reward initiative, that shift is noticed. And even if the prototype never ships as a product, the skills you develop, such as describing requirements clearly, iterating on feedback, and thinking about users, carry forward into any role.

Manus: The Powerful All-in-One Solution Autonomous Agent

Manus is different from the tools above. It is not a writing partner or an embedded assistant. It is an autonomous agent that takes a high-level goal, breaks it into steps, executes them, and adjusts as it goes. You describe what you want done, and Manus does it: researching, coding, building, and delivering a result that would normally require coordination across multiple people and systems.

I used Manus to build a proof-of-concept for a negotiation practice tool, as described in Chapter 7. I gave it the concept, the persona logic, and the experience design. It turned that input into a functional prototype, shifting the conversation from "imagine this product" to "here is the product." That speed mattered because there was a real internal deadline, and waiting for a traditional development cycle would have missed the window.

For professionals, Manus points to a future in which the ability to clearly describe what you want and evaluate what comes back becomes more important than the ability to execute every step yourself. Your role shifts from doing the work to designing it, setting the objectives, defining quality, and providing oversight. That transition is already underway in many organizations.

Manus is one of the tools making that future concrete.

Part Two: The Ten Systems Reshaping the Workday

The platforms above are the ones most professionals encounter first, but AI is being embedded into the workplace through many more points of contact. The ten systems below offer a broader map of where that is happening. Some are already widely used while others are just beginning to appear, but what connects them is the same underlying pattern: they reduce the time and coordination that certain categories of work once demanded.

1. Generative AI Platforms

For most professionals, generative AI platforms remain the starting point. Tools like ChatGPT, Claude, and Gemini allow users to interact with AI in natural language, asking questions, requesting drafts, or exploring ideas without needing specialized training. According to the Metrigy AI for Business Success 2025-26 study, employees who regularly use generative AI platforms save an average of 11.8 hours per week, roughly 29.4 percent of their working time. By early 2026, ChatGPT alone had reached approximately 900 million weekly active users, operating alongside other enterprise-grade models.

Research from the Stanford Digital Economy Lab supports this. Their study, Generative AI at Work, found that once organizations begin using AI tools, productivity gains appear quickly as routine drafting and summarization tasks are automated. The role of the professional moves toward reviewing, refining, and applying judgment. This does not reduce the importance of human thinking. It moves it to a different stage in the process.

2. No-Code and Low-Code Automation

Much of modern work still involves moving information from one system to another. A spreadsheet must trigger an approval. A contract signature must start an onboarding process. A report must be assembled from several data sources.

No-code and low-code automation tools address that layer of work. Platforms such as Zapier, Make, and Airtable allow non-technical employees to connect software systems and automate routine tasks without writing code. By 2026, Zapier alone supported nearly 8,000 application integrations, including more than 300 native AI services.

KPMG reports that automation can dramatically reduce the time spent on repetitive administrative tasks in functions such as finance, HR, and compliance, while also improving reliability and consistency. The value lies not only in time savings but also in reliability. Processes become more consistent when embedded in systems rather than relying on memory. Human resources teams automate onboarding sequences. Logistics operations automate disruption alerts. In both cases, coordination happens without manual follow-up.

3. Custom AI Agents and Workflow Builders

Instead of connecting existing tools, organizations are beginning to build entirely new workflows using AI agents. Platforms such as Manus, Lovable, and Replit Agent enable users to generate functional software prototypes from natural-language instructions.

This shift shortens the distance between idea and implementation. In traditional enterprise environments, many operational improvements never materialized because the effort required to build custom software was too large. AI-assisted development reduces that barrier. Healthcare organizations have begun exploring this approach with patient service workflows, with some hospitals reporting operational cost reductions of up to 40 percent.

The important point is structural. Professionals who previously could only request new software features can now participate in building the first version themselves.

4. AI Productivity Layers

Rather than interacting with AI through a separate interface, professionals increasingly encounter it embedded directly within their working environment. Microsoft Copilot and Notion AI are examples of this approach. These systems integrate AI assistance into documents, spreadsheets, notes, and collaboration platforms.

The benefit is not only efficiency but continuity. Modern work often requires constant switching between documents, communication platforms, and research tools. Embedded AI reduces context switching by bringing assistance directly into the workspace. A financial analyst asks Copilot to summarize historical performance data inside a spreadsheet. A project manager generates an executive briefing from meeting notes. The AI becomes a layer within the workflow rather than a destination outside it.

5. AI-Driven Learning Tools

Organizations generate vast amounts of knowledge, but much of it remains difficult to access. Policies, research reports, internal documentation, and training materials accumulate faster than employees can absorb them.

Platforms such as NotebookLM and Perplexity turn large collections of information into more usable formats. Instead of reading hundreds of pages sequentially, professionals can begin with an overview and drill into specific sections as needed. Corporate training teams use these tools to create customized learning paths tailored to different roles. Medical researchers use them to identify patterns across hundreds of dense clinical reports. Compliance teams use them to interpret new regulations and transform them into accessible training material.

The key advantage is speed of comprehension. Access to information is no longer the constraint. The challenge is to absorb and apply it quickly enough to remain effective.

6. Voice and Video Synthesis

Communication inside organizations increasingly relies on digital media. AI voice and video synthesis tools simplify production.

Platforms such as ElevenLabs support voice generation in more than 70 languages, while services like HeyGen and Synthesia produce customizable digital avatars that can deliver scripted content. For global organizations, training materials that once required separate recording sessions for each region can now be adapted quickly for multiple languages and audiences. The broader effect is not simply lower production cost. Communication becomes easier to update, translate, and distribute at scale.

7. AI Research and Knowledge Tools

Professional decision-making often depends on quickly gathering reliable information. Traditional search engines return lists of links that must be evaluated manually. AI research tools such as Perplexity, AlphaSense, and Elicit streamline that process by synthesizing information directly while providing clear source references. Investment banking analysts evaluating potential mergers can use generative-AI due diligence platforms to assemble background information in hours rather than days. Pharmaceutical companies examining patent landscapes before drug trials can rely on AI-powered patent and technical-intelligence tools such as Derwent, Patsnap, and Cypris to identify relevant constraints early in the process. The technology does not replace expert analysis; it accelerates the initial discovery stage, allowing professionals to spend more time evaluating implications.

8. Creative Generative AI

Creative generative AI now spans both experimental art tools and tightly governed enterprise systems. Tools like Midjourney and Flux have shown how non-specialists can create sophisticated, art-direction-level visuals from text prompts, dramatically lowering the barrier to early-stage visual ideation. Enterprise-oriented platforms such as Adobe Firefly, Getty Images' Generative AI, and Shutterstock's licensed-content models apply similar capabilities in a more controlled way, training on curated, rights-cleared image libraries and offering commercial-use safeguards to address corporate copyright concerns.

Marketing teams use these tools for concept development, producing campaign mock-ups and visual direction before formal design cycles begin, while architectural and design firms increasingly rely on them to turn sketches and massing studies into client-ready renderings. The technology does not eliminate the need for professional designers; instead, it shifts their focus to refining the strongest concepts, ensuring brand and legal compliance, and making higher-level creative decisions, rather than manually iterating every draft.

9. AI Scheduling and Task Management

Time has become one of the scarcest resources in professional life; meetings accumulate quickly, and priorities change frequently, leaving little uninterrupted space for focused work. AI scheduling applications such as Reclaim.ai and Motion analyze calendars and automatically rearrange meetings, tasks, and routines to preserve blocks of focused time. Early data from vendors and higher education studies suggest that AI scheduling and related automation tools can significantly reduce coordination and administrative overhead, with some vendor reports indicating reductions of up to 40 percent of the workweek.

Hospitals and clinical operations teams are beginning to use AI scheduling engines to manage complex staff rotations and on-call coverage, while global software organizations rely on these assistants, along with tools like Calendly and AI-enhanced collaboration platforms, to coordinate work across time zones without constant manual rescheduling. The issue these systems address is subtle but important: productivity depends not only on the tools people use to complete tasks, but also on whether their calendars are structured to support sustained attention in the first place.

10. Personal Knowledge Systems

The final system focuses on something less visible but equally important: professional memory.

Personal knowledge platforms such as Notion and Obsidian let individuals capture ideas, research notes, project insights, and references in structured formats for later retrieval. Over time, these collections become a searchable archive of accumulated experience. Without such systems, much professional insight disappears. Notes remain scattered across emails, documents, and notebooks. Valuable patterns are forgotten simply because they cannot be found when needed.

For many professionals, AI workspace tools have made this more accessible than dedicated platforms. Claude Projects, ChatGPT Projects, Perplexity's Spaces, and similar models let you store context, reference materials, and past outputs directly in the environment where you already work. You can return to a project weeks later and ask what was decided, what was left unresolved, or how a previous analysis was structured, and receive a useful answer without rebuilding the context from scratch.

Whether through a dedicated knowledge platform or an AI workspace, the principle is the same. Consultants track insights from previous engagements. Legal professionals maintain organized repositories of case references. Engineers document technical decisions for future projects. Each new piece of work draws on the accumulated context of earlier ones, reducing the time required to begin and improving the quality of what gets produced. That compounding effect is slow to notice and difficult to reverse once lost.

What This Means for You

The ten systems above show that AI is entering the workplace through many points of contact, not through a single wave. Some organizations start with writing tools. Others begin with automation, research assistants, or scheduling. In most cases, the first use case simply reveals the next opportunity, as teams notice where time is lost, where processes stall, and where knowledge becomes difficult to manage.

The most effective adoption strategies avoid introducing everything at once. Companies move faster when they focus on a few specific workflows and improve them step by step. The same principle applies to individuals. The goal is not to master every tool mentioned in this chapter. What matters is recognizing where these systems intersect with your own work and experimenting with them in ways that reduce the effort required by the tasks you perform most often.

Adaptation does not tend to arrive as a single transformation but develops gradually, drafts taking less time to produce, research becoming easier to synthesize, coordination requiring fewer steps, and knowledge that once disappeared into folders becoming easier to retrieve and reuse. Over time, those incremental improvements accumulate into a different work rhythm.

Reducing routine effort through technology expands what professionals can accomplish, freeing attention for judgment, context, and the decisions that require a human perspective. The people who adapt best will not be the ones chasing every new tool, but the ones who learn where these systems meaningfully strengthen the work they already do.

13

THE CHANGING LANDSCAPE

Every industry has its own version of the same story. The tools arrived, the work changed, and people sorted into those who moved with it and those who held on to the way things had always been done, until the gap between their approach and what the work now required became too wide to ignore. What differs from one field to the next is the details and the timing, not the underlying shape of what's happening.

This chapter examines how that story is unfolding across sectors. It doesn't cover everything (every field is more complex than any single chapter can hold), but it traces the patterns that show how AI is reshaping roles in ways that affect real people, whether they work in a hospital, a classroom, a courtroom, a marketing agency, or a warehouse. If your field isn't mentioned here, the pattern still applies, even if the specifics do not.

Healthcare: From Diagnosis to Decision Support

In healthcare, AI is reshaping the doctor's role rather than eliminating it, shifting attention from the administrative and diagnostic groundwork that once consumed entire days to the human judgment that no algorithm can replicate.

Diagnostic tools can now analyze medical images, flag potential conditions, and surface patterns across patient histories faster than any human team. A 2025 study by OpenAI and Penda Health, a clinic network in Kenya, found that doctors using an AI clinical copilot saw a 16 percent drop in diagnostic errors and a 13 percent reduction in treatment errors across nearly

40,000 consultations. The AI served as a second set of eyes running in the background, not making decisions but raising flags.

This means a change in role, not a loss of role, for healthcare professionals. Doctors spend less time on pattern recognition and more on patient communication, treatment planning, and the kind of judgment that requires understanding a person's full context, not just their test results.

Nurses find that AI helps with documentation and triage, freeing time for the hands-on care that patients remember. The most valued healthcare workers in the coming years will not be the ones who can process the most data. They will be the ones who can interpret what the data means for a specific patient sitting in front of them.

Education: The Teacher's Role Expands

In classrooms, AI is starting to handle much of what used to fill a teacher's evenings: grading, generating quizzes, adapting materials to different learning levels, and tracking student progress across assignments. Recent field studies of AI grading systems report reductions of around 80 percent in grading time, with teachers redirecting hundreds of hours per year toward planning and student support.

Platforms that adjust difficulty in real time based on how a student is performing, such as Squirrel AI's intelligent adaptive learning system, continuously assess what a learner understands, select the next problem or explanation, and provide teachers with dashboards that highlight where individual students or small groups are getting stuck. These systems are already deployed at scale in K–12 after–school programs and are beginning to appear in mainstream schools and universities to combine individualized practice with humanled discussion and coaching.

This does not make teachers less necessary. It makes them more important in the areas where they were always needed most: explaining difficult concepts in ways that connect to a student's experience, recognizing when a student is struggling emotionally rather than academically, designing group activities that build collaboration, and creating a learning environment where curiosity feels safe.

During my own master's program, I saw this firsthand. AI tools helped me manage the workload, but the moments that changed how I thought came from conversations with professors and classmates who challenged my assumptions and pushed me to connect ideas I would not have connected on my own. AI can deliver content. It cannot create the conditions that lead someone to care about what they are learning. That is still a human job.

Legal: From Document Review to Strategic Counsel

The legal profession was among the earliest knowledge work fields to encounter AI at scale. Contract review, case law research, due diligence, and regulatory analysis are all tasks where AI can process volume faster than any associate. Law firms that adopted AI tools early reported significant reductions in the time required for document-intensive work. Some legal departments report reductions of 80 to 90 percent in contract review time and substantial decreases in outside counsel spend after deploying AI-driven review platforms.

The entry-level tasks that used to serve as training grounds, such as reading thousands of pages of discovery, summarizing case precedents, and flagging contract deviations, are increasingly handled by AI. That does not eliminate the need for lawyers. It moves the value upward. The premium now goes to professionals who can advise clients on strategy, navigate ambiguity, manage relationships, and exercise judgment in situations where the law provides no clear answer.

For junior lawyers, this changes the way early career development looks. The path to seniority used to run through volume: hours spent in document review, building familiarity with how contracts and cases are constructed.

That work is largely automated now. What matters earlier in a career is the ability to frame problems clearly, communicate with precision, and think critically about what an AI-generated summary might be getting wrong or leaving out. The technical foundation still matters. But it is no longer enough on its own.

Finance: From Reporting to Interpretation

In finance, AI is compressing the cycle between data and decision. Quarterly reports that once took weeks to assemble can now be drafted in hours. Anomaly detection tools continuously scan transactions, flagging potential fraud or compliance issues in real time. In practice, major payment providers such as Stripe report scanning over 1,000 characteristics per transaction and cutting fraud with model responses around 100 milliseconds, while banks like Commonwealth Bank of Australia have documented roughly 30 percent reductions in fraud after deploying realtime AI detection. Forecasting models incorporate broader data sets than any analyst could manually review.

The professionals who thrive in this environment are not the ones who can build the most complex spreadsheet. They are the ones who can look at what the model produced and ask: Does this make sense given what I know about the business, the market, and the customer? Financial analysis is becoming less about production and more about interpretation. The human value lies in connecting numbers to narrative, explaining what the data means for the people who must make decisions based on it.

Marketing and Media: The Authenticity Premium

Marketing is the sector where AI's creative capabilities are most visible. Content that once required a photographer, a copywriter, a designer, and a video editor can now be produced by a small team using generative tools. A 2025 HubSpot report found that 55 percent of marketers already use AI as a core part of their content creation process.

The result is a market flooded with content. AI can generate polished copy, redesign layouts, and suggest hooks optimized for engagement. But research from 2025 found that campaigns developed with heavy AI assistance scored higher on technical execution and lower on distinctiveness. When teams relied too much on AI prompts, their ideas clustered around the same metaphors and visual tropes. Only when human creators set the direction and used AI as a production tool did the work regain originality.

This creates what some observers call the authenticity premium. Experimental studies show an 'AI-authorship effect': when consumers believe an emotional ad was written by AI rather than a human, they rate it as less authentic and report weaker purchase intentions, even when the content is otherwise identical. In an environment where anyone can produce professional-looking content, the scarce resource is not production quality. It is trust. Audiences can sense when something was made by someone who cares about the subject, versus when it was assembled by a tool. MIT researchers warned in 2024 that synthetic content would make it harder than ever to distinguish what is online. Recent reviews of consumer trust in AI-generated marketing argue that overuse of synthetic content triggers a measurable 'trust penalty' for brands, reinforcing the value of consistent human voices. That difficulty increases the value of voices that are consistent, verifiable, and grounded in genuine experience.

For marketing professionals, the career implication is this: the skills that matter most are not prompt engineering or tool fluency, though those help. They are taste, judgment, cultural awareness, and the ability to tell stories that feel real because they are.

Operations and Supply Chain: From Reactive to Predictive

In operations and supply chain management, AI is changing the work from reactive problem-solving to predictive coordination. Tools can now monitor inventory levels, track supplier performance, predict disruptions based on weather and shipping data, and recommend adjustments before problems escalate. Case studies of manufacturers adopting AI-based forecasting report forecast accuracy improvements of 25–30%, 20–30% reductions in excess inventory, and double-digit cuts in disruption-related losses within the first year.

This changes the daily work experience. Professionals who used to spend their weeks chasing updates, reconciling data across systems, and responding to surprises are now expected to spend that time on strategy: supplier development, value chain design, risk diversification, contingency planning, and cross-functional alignment. The operational details are increasingly handled by systems. The human role shifts toward designing those systems, setting their parameters, and intervening when the situation calls for judgment that no model can provide.

The Pattern Across All of Them

If you look across these sectors, the pattern is consistent. AI takes over repetitive, volume-driven, and pattern-dependent work. What remains, and what grows in importance, is the work that requires context, relationship, judgment, and the ability to act in situations where the right answer is not obvious.

This is not an easy transition. For many professionals, the tasks being automated are the ones they spent years mastering. Letting go of that mastery feels like losing ground, even when the new work is more meaningful. That emotional difficulty is real, and this book does not dismiss it.

But the evidence from every sector points in the same direction. The professionals who adapt are the ones who accept that their value is migrating, not disappearing. They learn the tools. They keep the judgment. And they focus their energy on the parts of work where being human is not a limitation but the entire point.

14

WHAT AI CANNOT OWN: JUDGMENT, TRUST, AND THE WORK THAT REMAINS HUMAN

There is a question running beneath everything in this book, even when it goes unspoken. If AI can draft contracts, summarize research, generate marketing campaigns, write code, analyze patient data, and build functional apps from a single sentence, what exactly is left for us?

The answer is more concrete than it might seem. AI cannot assume moral responsibility for a decision gone wrong. It cannot build the kind of trust that comes from being physically present when something matters. It cannot walk a junior colleague through a crisis of confidence or synthesize lived experience across multiple domains into an insight that has never existed before. And in the middle of a complex situation where the data is incomplete, and the stakes are personal, it cannot determine what the right thing to do is.

These are not soft skills on the margins of professional life. They are the hard architecture of how value is created across every organization, industry, and level. This chapter is about protecting and strengthening them.

Ethical Judgment: The Weight AI Cannot Carry

AI systems optimize. They correlate data points, flag anomalies, and recommend actions. What they cannot do is stand in the moral space where a decision means "this is right or wrong for this person, in this context, at this time."

A 2024 review of AI in medical research ethics found that, even when AI supports ethical review boards, the systems amplify biases embedded in training data, ranging from biased consent language to uneven privacy protections across population groups. The models encode what humans have done, not what humans ought to do.

In hiring, the pattern is similar. A late-2023 IBM-associated survey of more than 8,500 IT professionals found that 42% of companies were already using AI for recruiting and HR, and other surveys in the same period reported that majorities of employers were worried that these tools could introduce bias into hiring decisions.

Global economic institutions have warned that AI-driven decision-making in hiring and lending risks entrenching existing inequalities under a veneer of neutrality, as algorithmic systems often encode historical discrimination present in their training data. The lesson applies to every profession. AI can flag statistical risk. It cannot assume moral responsibility. Someone still must decide whether a vendor who defaulted under pressure deserves a second chance, whether a technically legal contract clause is morally defensible, or whether a cost-saving decision will erode trust with a critical partner. Those decisions cannot be delegated to a model. They must be owned by a person.

Creative Synthesis: What Machines Cannot Invent

Creativity in a professional context is about synthesis, or the ability to connect ideas from different domains into something that did not exist before.

A 2024 study in *Science Advances* examined how writers used GPT-4 to draft short stories. The findings were revealing. AI improved the output of less experienced writers by giving them better structure and more polished sentences. But for already-creative writers, AI made little difference in quality. More importantly, all AI-assisted stories became increasingly similar to one another. The machine, drawing from the same corpus, produced work that was more homogeneous, not more original.

A 2025 analysis of AI-assisted creative work in advertising and product design found the same pattern. Campaigns developed with AI tools scored higher on technical execution but lower on distinctiveness. When teams relied heavily on prompts, their ideas clustered around the same metaphors, structures, and visual patterns. Only when human creators set the direction and used AI as a production tool did the work regain its originality.

This applies far beyond advertising. In any field, the moment of creative value is when someone sees a connection that the data alone does not suggest. A nurse who draws on her experience with patient anxiety to redesign an intake process. A finance professional who applies lessons from behavioral economics to improve how risk is communicated to a board. A teacher who adapts a technique from improv theater to make statistics feel intuitive to a group of reluctant students. AI can assist once the connection has been made. But the connection itself, the leap across domains, remains human work.

Trust: The Currency That Cannot Be Automated

Trust is where AI's limits become most visible and most consequential. In 2025, the World Economic Forum warned that "AI will not wait for trust to catch up." Its message was clear: organizations that deploy AI without addressing trust should expect resistance, workarounds, and compliance failures from their own people.

Recent research on AI-assisted hiring shows that job seekers often view algorithmic screening as less fair than human review, even when the AI's recommendations are accurate, and that adding a human reviewer back into the loop significantly improves perceived fairness. When candidates were told that a human reviewer would still interpret the AI's output, trust increased significantly. The human presence was not a legal formality. It was a psychological anchor for fairness.

Consider a scenario described in recent supply-chain risk analyses: a European retailer deploys an AI-driven supplier risk tool that flags a long-time partner as 'high risk' after a spike in late deliveries during a port strike.

The model recommends freezing the contract, but a manager who knows the supplier challenges the recommendation, showing that the delays stem from external shocks rather than poor performance. AI can analyze behavior, flag anomalies, and suggest risk mitigation strategies. But it cannot reassure a supplier who fears being cut off. It cannot reassure a team that the technology is being used with them, not against them. It cannot sit across a table and say, "I understand the situation, and here is what I think we should do." That work is done by people who listen, explain, and stand behind their decisions.

Mentorship: The Relationship AI Cannot Replicate

Several studies between 2024 and 2026 explored AIdriven mentoring tools, chatbots that answer questions, suggest resources, and simulate coaching conversations, and concluded they are useful for onboarding and surfacing learning paths but cannot replace the emotional and relational dimensions of mentoring.

A 2025 review of AI and empathy in caring relationships notes that while AI-generated responses can sound empathetic, users often detect their artificial nature, undermining perceived authenticity and reducing trust and connection. Researchers caution that using bots for mentoring can improve access to information, but does nothing to build the personal networks through which opportunities flow, and surveys suggest that around half of workers still land roles through personal or professional connections. Mentees reported feeling that they were talking to a welltrained script, not a person invested in their growth.

Mentoring lives in the space where knowledge runs out and judgment takes over, helping someone navigate what no framework can fully prepare them for. A human mentor can watch a colleague hesitate during a difficult call and later say, "You held the line well, but next time let me show you how to soften the ask without giving ground." An AI tool might suggest a script. It cannot read the flicker of doubt in someone's expression or the hesitation in their voice.

Recent work on AImediated relationships finds that chatbots can prompt reflection and even foster a sense of closeness in controlled interactions, but that perceived authenticity and a sense of being genuinely "known" remain lower than with human partners, especially when people are aware they are talking to AI. In mentoring contexts, psychologists argue that while AI can scaffold reflection and provide feedback, it cannot provide the belonging, sponsorship, and advocacy that arise from real human mentors willing to stake their own reputations on someone's growth.

A growing body of research on AI companions and support tools concludes that while AI can nudge reflection and offer advice, it cannot substitute for the sense of belonging, sponsorship, and advocacy that comes from a human mentor.

The [likely] Future Is Augmentation

Adapter Profile: Roberto, The Displaced Manager Who Found a Second Act

Roberto managed twelve account coordinators at a logistics firm in Miami. He was fifty-three and had been there seventeen years. Then his company restructured. An AI platform replaced most of his team's coordination work. His team was cut to four. Roberto was offered early retirement.

The first two months were difficult. He described them as "waking up with no reason to check my phone." He looked at job postings and found that most required AI skills he did not have.

What started the process was a conversation with his daughter, a college junior. She asked him to describe what he did every day, not his title, but the specific decisions. After an hour, she showed him the list: conflict resolution, client de-escalation, cross-cultural communication, team motivation, vendor negotiation, and the ability to read a situation before the data caught up. "Dad," she said, "none of that is in the AI platform."

Roberto started mentoring through a workforce development program. He discovered that his decades of relational knowledge were exactly what the next generation lacked. Within six months, he was consulting for three import-export firms on client relationships and human skills.

He earns less than before. He works fewer hours. And he will tell you he has never felt more useful. AI can process information. It cannot sit across from a frustrated client and say, "I understand. Let me tell you what we are going to do." Well, technically, it can, but not in the same fashion as a human-to-human interaction.

Between 2024 and 2026, the dominant narrative moved from 'Will AI replace humans?' to 'How can humans and AI work together effectively?' Studies on AI-augmented teams consistently show that the best results come when AI handles the routine, data-intensive work while humans focus on interpretation, relationship, and judgment.

That means using AI to monitor performance, flag anomalies, and draft standard outputs, while humans focus on strategy, governance, stakeholder alignment, and the relational work that holds organizations together when things get complicated.

The key insight is simple. AI is not a competitor for human roles. It complements human judgment. The people who thrive in the AI era will not be those who fear the tools or those who hand everything to them. They will be the ones who learn to use AI for what it does well and protect the skills that remain stubbornly, structurally, and productively human.

Your judgment. Your relationships. Your ability to see beyond the data. Those are not relics of a pre-AI world. They are the foundation of what comes next.

15

WHY AI ADOPTION FAILS AND HOW TO BUILD A TEAM THAT GETS IT RIGHT

The Gap Nobody Talks About

Every organization I have worked with or spoken to in the past three years tells a version of the same story. Leadership announces an AI initiative. Enthusiasm is high. Employees attend a launch event or an introductory workshop. There is genuine excitement. People can see the potential. And then, within weeks, almost nothing changes.

The tools are available. The training has been delivered. The strategy deck exists. But the daily work looks the same as it did before the initiative launched. Emails are still written the same way. Reports are still assembled manually. Meetings still run without AI-assisted summaries. The adoption curve that looked so promising in the first week flattens into a line that barely moves.

This is the pattern I saw at Unilever, and it is the same pattern I have seen in conversations with professionals across industries and company sizes. The problem is almost never the technology. The tools work. The problem is the space between the tool and the person using it. That space is filled with habits, fear, unclear expectations, and a lack of support that no product demo can solve.

If you are a leader responsible for bringing AI into your organization, or if you are a professional trying to understand why your team has not moved despite a clear opportunity, this chapter is the most practical one in the book. It covers what goes wrong and, more importantly, what to do about it.

The Five Adoption Failures

After running workshops for thousands of people, leading internal AI initiatives, conducting research for this book, and observing how multiple teams responded to the same tools, I began to see the same failure modes repeat themselves. They are not random. They follow a pattern we explore next.

Failure 1: Starting with the tool instead of the problem

The most common mistake is introducing AI as a technology rather than as a solution to a specific problem. When organizations lead with what a tool can do, employees are impressed but unsure what to do next. A demo that looks compelling in a conference room almost never translates into changed behavior the next day.

The fix is to reverse the sequence. Start by identifying the three to five tasks that consume the most time in each role, the ones that repeat weekly and add limited strategic value. Then ask: can AI reduce the time or effort required for any of these? When the entry point is a real frustration rather than a feature list, adoption has a reason to last.

At Unilever, the workshops that worked best were those in which I interviewed participants beforehand to understand their specific challenges. A procurement analyst drowning in spend reports needs a different demonstration than a marketing coordinator managing campaign approvals. Same tool, different entry point. That specificity made all the difference.

Failure 2: Treating AI adoption as a one-time event

A single workshop, no matter how well designed, does not change behavior. It creates awareness. Awareness is the first step of the ADAPT method, not the last. Without follow-up, reinforcement, and visible examples of colleagues successfully using AI, the initial enthusiasm fades within days.

The fix is to design adoption as a rhythm rather than an event. A practical structure I found effective was a four-week cycle. Week one: introduce the tool on one real task. Week two: refine the workflow and document the steps. Week three: Share what you learned with one colleague. Week four: decide whether the workflow stays or goes. That cycle, repeated across a team, creates momentum that a single training session cannot.

Failure 3: Security guardrails that block the learning moment

This one is difficult because the guardrails exist for good reasons. Large organizations cannot afford to have sensitive data accidentally shared with external AI systems. The challenge is that overly restrictive environments often keep employees from experiencing the moment when AI becomes genuinely useful. If the tool can only do basic tasks because most features are locked down, people walk away thinking AI is not ready yet. The problem is not the technology. It is the environment.

The fix is not to remove guardrails but to create safe practice spaces. Organizations can set up sandbox environments with non-sensitive data where employees can experiment freely. They can identify low-risk workflows where AI can be used without triggering compliance concerns. The goal is to let people experience the "this actually helps" moment within the organization's required boundaries. Without that moment, adoption is going to be difficult to manage.

Failure 4: No visible champions inside the organization

People adopt tools they see working for someone they trust. When AI is driven only by IT or consultants, it feels abstract. It needs internal faces, like the colleague who cut reporting time in half with Copilot or the manager who shares a prompt that saved hours.

The fix is a champions network: a small group of curious early adopters across functions, given time, visibility, and permission to experiment. These are not AI experts; they are just practitioners. At Unilever (Chapter 3), this approach scaled from a handful to hundreds of people in months, not through technical training but by enabling experimentation and sharing results.

Failure 5: Measuring the wrong things

Many organizations measure AI adoption by counting activated licenses, recorded logins, or completed training sessions. These metrics tell you who has access. They do not tell you who is working differently.

The fix is to measure behavioral change, not tool usage. Better questions include: How many workflows have been redesigned to include AI? How much time has been saved on specific recurring tasks? How many people have shared a working AI workflow with a colleague? These are harder to count, but they reflect what matters: whether the work has changed.

Building an AI-Ready Team

Fixing the failures above likely addresses the immediate adoption problems (or most of them). But lasting change requires something deeper. It requires building a team that can continuously absorb new tools and methods, not just this year's AI rollout but next year's as well, and the one after that.

An AI-ready team is not a team of technologists. It is a team that has developed four specific capabilities.

Capability 1: Experimentation as a habit

Teams that adopt AI well are teams that already know how to run small experiments. They test an idea on a limited scope, evaluate the result, and decide whether to continue or stop. This does not require a formal innovation program. It requires a culture that accepts trying something new, even if it does not work.

If your team does not have this habit, start by carving out protected time. Even one hour per week dedicated to testing something new can change the dynamic over a few months. The key is that the experiments are real. They use actual work tasks, not hypothetical scenarios.

Capability 2: Workflow literacy

Most professionals can describe what they do, but not how they do it. They cannot break their work into steps, identify bottlenecks, or articulate which steps are repetitive and which require judgment. That kind of workflow literacy is essential for AI adoption, because AI works best when you can point it at a specific, well-defined task rather than a vague objective.

Building this capability does not require process mapping software or consulting engagements. It requires asking your team a simple question: "Walk me through exactly what you do between receiving this request and delivering the result." The answer almost always reveals two or three steps where AI could help, and at least one step that the team did not realize was consuming so much time.

Capability 3: Prompt fluency

Learning to work with AI tools is less a technical discipline than an extension of a skill most professionals already have: the ability to communicate clearly, precisely, and with a specific outcome in mind. The professionals who get the most value from AI are the ones who can describe what they want in specific, structured terms: the context, the audience, the format, and the constraints.

Teams build this fluency fastest through sharing. When one person discovers a prompt that works well for a common task and shares it in a team channel or a simple document, the entire team benefits. Over time, the team accumulates a prompt library that becomes a practical asset, a collection of tested patterns that new members can use from day one.

Capability 4: A learning loop between doing and teaching

The teams that sustain AI adoption are the ones where learning flows in both directions. Junior team members often discover new tools and techniques faster. Senior members bring context and judgment about where those tools create real value. When both groups share openly, the team moves faster than any individual could.

The simplest structure for this is a recurring short session, fifteen to thirty minutes, biweekly or monthly, where team members share one thing they tried, what happened, and what they learned. No slides required. No formal agenda. Just truthfully reporting on experiments, including the ones that did not work. Over time, these sessions build a shared vocabulary and confidence that make the next adoption cycle easier.

The Organizational Playbook

If you are leading AI adoption for a team, department, or entire organization, the principles above can be organized into a practical sequence.

Month 1: Diagnose. Interview ten to fifteen people across functions. Ask them what tasks consume the most time, where they feel stuck, and what they have already tried with AI. Listen for patterns. The answers will tell you where to start better than any vendor pitch.

Month 2: Pilot. Select two to three workflows where the pain is real and the risk is low. Pair each workflow with a tool, assign a small team, and set a clear success measure: time saved, quality improved, or adoption by others.

Month 3: Amplify. Share the results from the pilot, including what did not work. Identify the people who showed the most initiative during the pilot and invite them into a champions network. Give them visibility, not extra work. Their job is to make the learning contagious.

Month 4 and beyond: Sustain. Establish the learning loop. Run the biweekly sharing sessions. Track behavioral metrics, not just logins. Expand the pilots to new workflows based on what the first round taught you. Revisit the available tools quarterly, as tools will change and your team needs to adapt.

This is not a twelve-month transformation program. It is a repeatable cycle that gets faster each time your team runs through it. The first cycle is the hardest because you are building the habit of experimentation alongside the actual adoption. After the first cycle, the habit exists, and each subsequent wave of technology finds a team that already knows how to absorb it.

Why This Matters for Careers

For the individual professional, this chapter contains a career opportunity that is easy to miss.

Most organizations are struggling with exactly the problems described above. They have invested in AI tools, but cannot get their teams to use them. They have run workshops that generated excitement and nothing else. They are looking for people who can bridge the gap between technology and behavior change.

If you understand why adoption fails and how to build teams that adopt well, you possess a skill in high demand and short supply. That skill does not require a technical background. It requires the ability to listen, to design practical experiments, to teach without condescension, and to measure what matters.

Whether you apply that skill within your current organization as an internal champion or outside it as a consultant, advisor, or trainer, the need is real and growing. The ADAPT method equips you to navigate your own career. This chapter equips you to help others navigate theirs. In the AI era, that combination is worth more than any single technical certification.

16

THE ADAPTER'S PROMISE

The Scouts Return

In the Preface, I told you about the honeybee colony.

When the environment changes, whether through a drop in temperature, the loss of a food source, or an unfamiliar threat, the hive does not hold a meeting. It does not commission a report. It sends scouts, the worker bees that leave the colony, explore new territory, and return with information. They communicate through the waggle dance, encoding direction, distance, and quality into a language the colony can read. The hive does not wait for certainty. It follows the scouts.

This book has been my waggle dance. I went first, not because I was braver or smarter, but because the signals were impossible to ignore and my restlessness would not let me stand still. I tested the tools. I failed in public. I learned to separate what AI could do from what it could not, and I discovered that the experience I thought was becoming obsolete was what made the tools useful.

Now the scouts have done their work. The question is no longer whether the environment has changed. The question is whether the hive will move. That question is for you.

What I Know Now

If I could distill everything in this book into a handful of truths, not the framework, not the method, but the raw lessons that remained after the writing was done, they would be these:

Stability can be misleading. The years I spent on a plateau taught me that the most insidious threat to a career is not disruption, but the absence of it. When things are stable enough, you stop asking whether they could be better. You mistake repetition for expertise. Failure forces you to examine your assumptions. Familiarity lets you sleep right through the alarm.

AI Exposes the Professionals Who Already Stopped Thinking. The professionals most threatened by AI are not those with outdated technical skills. They are the ones who stopped developing judgment. AI did not create that problem. It made it visible. If the work has been running on habit rather than active thinking for some time, that will show. But for professionals who have maintained the practice of asking why things work the way they do, the same tools become something different: a way to move faster on the thinking they were already doing. The difference in outcomes between those two groups will not come from the tools they use. It will come from what they brought to the tools before they ever opened them.

Your career is not a ladder. It is a portfolio of experiments. The professionals who adapt best have careers that look like investment portfolios: diversified, occasionally rebalanced, measured by total value rather than a single peak. Every role, every side project, every skill developed out of curiosity rather than obligation, is not a distraction from your career. They are your career.

The people who adapt are not smarter. They are braver about being beginners. Reinvention requires you to be bad at something for a while. That became clear when I first tried to build an app and again when teaching technology workshops, which I was still learning firsthand. Every time, the unease passed. The new skill did not. The willingness to look uncertain in pursuit of lasting capability is the single most important trait of an Adapter. No one adapts alone. My reinvention did not happen in isolation. It happened because Unilever gave me room to experiment, because colleagues shared what they were learning, because my family absorbed the late nights, and because professionals like Clara, James, Amara,

David, Mei, Roberto, and many others showed me that adaptation is a community practice. The scout does not survive without the hive. And the hive does not move without the scout.

Your First Seven Days

If the ADAPT method is the architecture, consider this your first brick. Seven days. One real task from your actual work. No grand plans, no expensive courses, no waiting for permission. Just one focused experiment that takes you from reading about AI to working with it.

Day 1: Name the Resistance. Identify one recurring task in your work that feels slow, repetitive, or disproportionate to its value. Do not try to solve it yet. Just name it and write it down. Naming the resistance is the beginning of Awareness.

Day 2: Pick one Tool. Based on the task you identified, choose one AI tool available to you today. Your only goal is to open the tool and ask it one question related to the problem you are trying to solve.

Day 3: Run a narrow Test. Use the tool on a small piece of your real task. Do not try to automate the whole process. Just aim to cut one part of it in half. Compare the output with how you would have done it without the tool.

Day 4: Measure the Change. Was it faster? Was the quality better or worse? What surprised you? Write down one sentence that captures the result.

Day 5: Document the Play. Write down the exact steps you took, including the prompt you used. Keep it simple, as if explaining to a colleague over coffee. This is your first "play."

Day 6: Teach it Once. Find one trusted colleague and spend five minutes showing them your play. The act of explaining will sharpen your own thinking, and their reaction will tell you something useful about how your organization relates to change.

Day 7: Decide What is Next. Did the experiment save time or create more work? Is it worth repeating? Decide whether to keep this play, refine it, or try a different task. Either way, you have just completed your first adaptation loop.

After seven days, you will have more practical knowledge about AI in your specific work than most professionals accumulate in six months of reading articles. You will have evidence you generated yourself. And you will have the lived experience of moving before you felt ready.

The Cemetery, Revisited

I want to end where I began. Not in Orlando. Not in the boardroom. Not at a screen. I want to go back to my hometown, Guarulhos.

A seven-year-old boy stands on a corner outside a cemetery, watching adults come and go. He has no resources, no connections, no plan beyond this: earn enough coins to buy a radio so he can follow his football team's matches. The work is simple. Watch the cars, collect the payment, endure the indifference. Some people see him and look away. Some ask questions. A few hand him coins and something small: a lollipop, a nod, a moment of recognition.

He does not know it yet, but he is learning the lesson that will define the next four decades of his life. When the world gives you nothing, you find a way forward. You observe. You approach. You negotiate. You adapt. You do not wait for the conditions to be perfect, because they never will be. You start with what you have, where you are, and you build.

That boy is still in me. He is in every chapter of this book. He is a teenager who talked his way past a knife on a bus by offering bubble gum. He is an intern who learned DOS because it was the only way forward. He is the procurement professional who sat in an office in Switzerland, questioning whether the career he had built was the one he still wanted to be building. He is a man in his late forties who stared at a screen one night in Florida and decided, once again, that movement was better than certainty.

And he is in you, too. Whatever your starting point looked like, whatever your version of that first small act of figuring it out was, the instinct underneath it is the same. The refusal to accept that this is as far as you go. The willingness to try something before you know how it ends. The belief that what you have lived through is not a weight but a foundation. That instinct is the essence of adaptation. It always has been.

The algorithm changed everything. But the Adapter changed first.

Welcome to the journey. I will see you on the platform.

QUICK NAVIGATION GUIDE

Use this guide when you want to return to a specific idea, exercise, or moment in your own adaptation journey. ADAPTERS is organized in three parts: the personal story, the ADAPT framework, and the practical future of work.

If you are feeling behind or uncertain, read the Introduction, then Chapter 1: When Everything Changed. These sections explain why AI changes the meaning of experience without erasing its value.

If your career looks successful but feels stalled, read Chapter 2: The Hidden Plateau. This chapter is for the reader who is performing well but no longer growing.

If you are rebuilding after failure or disruption, read Chapter 3: Starting Again, Starting Different. This is the comeback chapter, and the bridge from memoir to method.

If your career path feels scattered, read Chapter 4: Mastering the Polymath Approach. This chapter reframes variety as an advantage rather than a weakness.

To work through the ADAPT method, read Chapters 5 through 9 in sequence: Awareness: Notice what is changing before it becomes urgent. Direction: Choose a path without waiting for certainty. Action: Run small experiments and create evidence. Positioning: Make your growth visible and credible. Tech Empowerment: Use tools without becoming overwhelmed. The book defines

Part II presents a practical framework, with exercises and practices readers can begin right away.

For practical AI use cases, read Chapter 10: The Adapter›s Toolkit and Chapter 12: The Tools That Are Changing How We Work. These chapters show what AI-supported work looks like in daily practice.

If you are worried about AI mistakes, read Chapter 11: When AI Gets It Wrong. This is the guardrail chapter.

To understand what remains human, read Chapter 14: What AI Cannot Own. This chapter focuses on judgment, trust, and the human work that still matters most.

If you lead a team, read Chapter 15: Why AI Adoption Fails and How to Build a Team That Gets It Right. Use it when moving from personal adaptation to organizational adoption.

When you are ready to begin, read Chapter 16: The Adapter›s Promise and complete the seven-day challenge. The final chapter closes with a simple adaptation loop: test one AI use case, teach it once, decide what comes next, and keep moving.

NOTES

1. ReYOUniverse. (2024, July 21). A civilization 100 million years old [Video]. YouTube. https://www.youtube.com/watch?v=H7H3zVGXQ3k

2. Unilever. (n.d.). Unilever [Company website]. https://www.unilever.com

3. OpenAI. (2024). ChatGPT (GPT-4o version) [Large language model]. https://chat.openai.com

4. Google. (2024). Gemini (Gemini 2.0 version) [Large language model]. Google DeepMind. https://gemini.google.com

5. Anthropic. (2024). Claude (Claude 3.5 Sonnet version) [Large language model]. https://claude.ai

6. Lovable. (2024). Lovable [AI-powered web application builder]. Lovable Technologies. https://lovable.dev

7. Veit, F. (2026). Adapterslab.ai [Web platform]. https://adapterslab.ai

8. Manus. (2026). Manus [Agentic AI platform]. https://manus.im

9. Burnett, M., Hartmann, J., & Cain, J. (Executive Producers). (2009–present). Shark Tank [TV series]. ABC; Sony Pictures Television. https://abc.com/show/shark-tank

10. Gallup. (2026, January 24). Frequent use of AI in the workplace continued to rise in Q4. Gallup. https://www.gallup.com/workplace/701195/frequent-workplace-continued-rise.aspx

11. Microsoft. (2025, June 3). Microsoft's 2025 Work Trend Index Report reveals the rise of the Frontier Firm [News release]. Microsoft News. https://news.microsoft.com

12. Pew Research Center. (2025, February 24). U.S. workers are more worried than hopeful about the future use of AI in the workplace. https://www.pewresearch.org/social-trends/2025/02/25/u-s-workers-are-more-worried-than-hopeful-about-future-ai-use-in-the-workplace/

CHAPTER 1 - When Everything Changed

13. Roose, K. (2022, December 10). The brilliance and weirdness of ChatGPT. The New York Times. https://www.nytimes.com/the-brilliance-and-weirdness-of-ChatGPT

14. Satariano, A., & Metz, C. (2023, March 5). Using A.I. to detect breast cancer that doctors miss. The New York Times. https://www.nytimes.com/2023/03/05/technology/artificial-intelligence-breast-cancer-detection.html

Note. This article reports real-world cases in which artificial intelligence systems identify possible cancers on imaging that human radiologists initially overlook, illustrating how AI can flag critical findings that might be missed in clinical practice.

15. Piller, E. (2023). The ethics of (non)disclosure: Large language models in professional, nonacademic writing contexts. Rupkatha Journal on Interdisciplinary Studies in Humanities, 15(4). https://doi.org/10.21659/rupkatha.v15n4.01

Note. This article examines how professional writers in commercial and nonacademic settings employ large language models to generate and refine client-facing text, showing that AI tools are already integrated into copywriting and related creative workflows.

16. Niszczota, P., & Abbas, S. (2023). Insights from financial literacy tests of GPT and a preliminary test of how people use it as a source of advice (arXiv No. 2309.00649). arXiv. https://arxiv.org/abs/2309.00649

Note. This preprint reports that GPT models demonstrate high performance on financial literacy tests and are used by individuals as a source of financial advice, indicating that AI systems can produce analyses comparable to those of human financial experts.

17. Wang, G., & Ling, D. (2024). Fund performance driven by ChatGPT: Evidence from Chinese stock funds (SSRN Working Paper No. 4839977). SSRN. https://papers.ssrn.com/sol3/papers.cfm?abstract_id=4839977

Note. This working paper studies the relationship between ChatGPT-related developments and mutual fund performance, suggesting that AI-driven information and tools are influencing investment decisions and the behavior of financial markets.

18. Sociedade Esportiva Palmeiras. (n.d.). Home [Website]. Palmeiras. https://www.palmeiras.com.br

19. Securit S.A. (n.d.). Securit S.A. [Company website]. https://www.securit.com.br

20. Folha de S. Paulo. (n.d.). Folha de S. Paulo [News outlet]. Grupo Folha. https://www.folha.uol.com.br

21. Olivetti. (n.d.). Olivetti [Company website]. https://www.olivetti.com/en

22. Xerox. (n.d.). Xerox [Company website]. https://www.xerox.com/en-us

23. Grupo Eurodata Brasil. (n.d.). Grupo Eurodata Brasil [LinkedIn profile]. LinkedIn. https://www.linkedin.com/company/grupoeurodata/

24. DuPont. (n.d.). DuPont [Company website]. https://www.dupont.com

25. IBM. (n.d.). IBM [Company website]. https://www.ibm.com

26. SAP. (n.d.). SAP [Enterprise software]. https://www.sap.com

27. Laporte plc. (n.d.). Laporte plc [Wikipedia entry]. https://en.wikipedia.org/wiki/Laporte_plc

28. Evonik Industries. (n.d.). Evonik Industries [Company website]. https://www.evonik.com/en.html

CHAPTER 2 - The Hidden Plateau

29. Noh, H., Cho, Y., & Kim, S. (2024). The impact of career plateaus on job performance: A moderated mediation model. Behavioral Sciences, 14(2), 1–17. https://doi.org/10.3390/bs14020123

Note. This study defines career plateau as a perceived stagnation point in one's career and finds that plateaued employees show lower work engagement and in-role performance, as well as gradual psychological disengagement from their organization over time.

30. Li, X., Zhang, Y., & Chen, L. (2025). Career plateau and career withdrawal intentions among Chinese college physical education teachers: The mediating role of job satisfaction and moderating role of organizational commitmenthttps://www.frontiersin.org/journals/psychology/articles//full

Note. This article shows that career plateau significantly reduces job satisfaction and motivation to remain in the profession, and increases intentions to withdraw from one's career, highlighting the psychological costs of prolonged stagnation even when workers remain technically capable.

31. HP Inc. (n.d.). HP [Company website]. https://www.hp.com/us-en/home.html

32. Uber Technologies. (n.d.). Uber [Company website]. https://www.uber.com

33. Airbnb. (n.d.). Airbnb [Company website]. https://www.airbnb.com Reference to IBM Watson's Jeopardy! victory, 2011, p. 35]

34. Krizhevsky, A., Sutskever, I., & Hinton, G. E. (2012). ImageNet classification with deep convolutional neural networks. Advances in Neural Information Processing Systems, 25, 1097–1105.

CHAPTER 3 - Starting Again, Starting Different

35. Unilever. (2010). Unilever Sustainable Living Plan. Unilever. https://www.unilever.com

36. MIT Deshpande Center for Technological Innovation. (n.d.). MIT Deshpande Center. Massachusetts Institute of Technology. https://deshpande.mit.edu

37. Algramo. (n.d.). Algramo: Refill on the go [Company website]. https://www.algramo.com

38. TerraCycle. (n.d.). TerraCycle: Recycle the unrecyclable [Company website]. https://www.terracycle.com

39. The Open University. (n.d.). The Open University [University website]. https://www.open.ac.uk

40. Rover Group. (n.d.). Rover: Dog sitters and dog walkers [App]. https://www.rover.com

41. Whole Foods Market. (n.d.). Whole Foods Market [Company website]. Amazon. https://www.wholefoodsmarket.com

42. Ibarra, H. (2003). Working Identity: Unconventional strategies for reinventing your career. Harvard Business School Press.

43. The Farmer's Dog. (n.d.). The Farmer's Dog: Human-grade fresh dog food delivery [Company website]. https://www.thefarmersdog.com

44. Ollie. (n.d.). Ollie: Fresh dog food delivered [Company website]. https://www.myollie.com

45. OtterBox. (n.d.). OtterBox: Protective phone cases and accessories [Company website]. Otter Products. https://www.otterbox.com

46. University of Wisconsin–Madison. (n.d.). University of Wisconsin–Madison [University website]. https://www.wisc.edu

47. Ghosal, A. (2021, August 5). Canada's mustard seed crisis: Drought devastates crop, pushing prices to record highs. CBC News.

48. Perplexity AI. (2024). Perplexity AI [AI-powered search engine]. https://www.perplexity.ai

49. Google. (2024). NotebookLM [AI-powered notebook]. https://notebooklm.google

CHAPTER 4 - Mastering the Polymath Approach

50. Quesada, A. (2024, August 9). Cultivating the modern polymath with AI. Psychology Today. https://www.psychologytoday.com/us/blog/the-digital-self/202408/cultivating-the-modern- polymath-with-ai

51. Instagram. (n.d.). Instagram [Social media platform]. Meta Platforms. https://www.instagram.com

52. LinkedIn Corporation. (n.d.). LinkedIn [Professional networking platform]. Microsoft. https://www.linkedin.com

53. Notion Labs. (n.d.). Notion [Productivity and note-taking platform]. https://www.notion.so

54. Obsidian. (n.d.). Obsidian [Knowledge management and note-taking application]. https://obsidian.md

55. Figma. (n.d.). Figma [Collaborative design platform].https://www.figma.com

56 .YouTube. (n.d.). YouTube [Video sharing platform]. Google. https://www.youtube.com

57. Canva. (n.d.). Canva [Online design platform]. https://www.canva.com

58. Airtable. (n.d.). Airtable [Cloud collaboration platform]. https://www.airtable.com

59. European Parliament and Council of the European Union. (2024). Regulation (EU) 2024/1689 laying down harmonized rules on artificial intelligence (Artificial Intelligence Act). Official Journal of the European Union.

CHAPTER 5 - Awareness

60. Wuhan Municipal Health Commission. (2019, December 31). Report of clustering pneumonia of unknown etiology in Wuhan City [Press release].

61. Microsoft. (2024). Microsoft Copilot [AI assistant]. https://copilot.microsoft.com

62. Feedly. (n.d.). Feedly: Track the topics and trends that matter to you [RSS reader and AI research tool]. https://feedly.com

63. MIT Technology Review. (n.d.). MIT Technology Review [Technology publication]. Massachusetts Institute of Technology. https://www.technologyreview.com

64. Harvard Business Review. (n.d.). Harvard Business Review [Management publication]. Harvard Business Publishing. https://hbr.org

65. McKinsey & Company. (n.d.). McKinsey & Company [Management consulting]. https://www.mckinsey.com

66. CB Insights. (n.d.). CB Insights [Technology market intelligence]. https://www.cbinsights.com

CHAPTER 6 - Direction

67. eAdams, C., Pente, P., Lemermeyer, G., & Rockwell, G. (2023). Artificial intelligence and teachers' professional identity: Mapping the evolution of AI in education toward a co-adaptive and symbiotic future. AI & Society.

68. Schwartz, B. (2004). The paradox of choice: Why more is less. Ecco/HarperCollins.

69. Gilbert, D. T., & Ebert, J. E. J. (2002). Decisions and revisions: The affective forecasting of changeable outcomes. Journal of Personality and Social Psychology, 82(4), 503–514. https://doi.org/10.1037/0022-3514.82.4.503

70. Ericsson, K. A., Krampe, R. T., & Tesch-Römer, C. (1993). The role of deliberate practice in the acquisition of expert performance. Psychological Review, 100(3), 363–406. https://doi.org/10.1037/0033-295X.100.3.363

71. Clear, J. (2018). Atomic habits: An easy and proven way to build good habits and break bad ones. Avery/Penguin Random House.

72. Koh, A. W. L., Lee, S. C., & Lim, S. W. H. (2018). The learning benefits of teaching: A retrieval practice hypothesis. Applied Cognitive Psychology, 32(3), 401–410.

CHAPTER 7 - Action

73. Kolb, D. A. (1984). Experiential learning: Experience as the source of learning and development. Prentice-Hall.

74. Ries, E. (2011). The lean startup: How today's entrepreneurs use continuous innovation to create radically successful businesses. Crown Business.

75. Bandura, A. (1997). Self-efficacy: The exercise of control. W. H. Freeman.

CHAPTER 8 - Positioning

76. Pfeffer, J. (2010). Power: Why some people have it, and others do not. HarperBusiness.

CHAPTER 9 - Tech Empowerment

77. Mollick, E. (2024). Co-intelligence: Living and working with AI. Portfolio/ Penguin.

78. Noy, S., & Zhang, W. (2023). Experimental evidence on the productivity effects of generative artificial intelligence. Science, 381(6654), 187–192. https://doi. org/10.1126/science.adh2586

79. National Academies of Sciences, Engineering, and Medicine. (2025). Artificial intelligence and the future of work. The National Academies Press. https:// doi.org/10.17226/27644

CHAPTER 10 - The Adapter's Toolkit: A Professional Life in Practice

80. Di Stefano, G., Gino, F., Pisano, G. P., & Staats, B. R. (in press). Learning by thinking: The role of reflection in individual learning. Management Science. https://doi.org/10.2139/ssrn.2414478

CHAPTER 11 - When AI Gets It Wrong

81. Bohannon, M. (2023, June 8). Lawyer used ChatGPT in court - and cited fake cases. A judge is considering sanctions. Forbes. https://www.forbes.com

82. Harvard Kennedy School. (2024, December 5). GPTs and hallucination. Harvard Kennedy School – Ash Center for Democratic Governance and Innovation. https://www.hks.harvard.edu/publications/gpts-and-hallucination

83. IBM Institute for Business Value. (2023, August 13). New IBM study reveals how AI is changing work and what HR leaders should do about it. IBM. https://www.ibm.com/think/insights/new-ibm- study-reveals-how-ai-is- changing-work-and-what-hr-leaders-should-do-about-it

84. Stanford Institute for Human-Centered Artificial Intelligence. (2025, February 5). AI's fairness problem: When treating everyone the same is the wrong approach. Stanford HAI. https://hai.stanford.edu/news/ais-fairness-problem- when-treating-everyone-same-wrong-approach

85. Harvard Gazette. (2025, November 12). Is AI dulling our minds? Harvard University. https://news.harvard.edu/gazette/story/2025/11/is-ai-dulling-our-minds

CHAPTER 12 - The Tools That Are Changing How We Work

86. Microsoft. (2025). New employee Copilot usage: How interns adopt AI at work (Research report). Microsoft Research. https://www.microsoft.com/en-us/research/wp-content/uploads/2025/04/New-employee-Copilot-usage.pdf

87. Metrigy. (2025). AI for Business Success 2025–26 (Global research study). Metrigy.

88. Brynjolfsson, E., Li, D., & Raymond, L. R. (2025). Generative AI at work. The Quarterly Journal of Economics, 140(2), 889–936. https://doi.org/10.1093/qje/qjaf002

89. Zapier. (2026). Power your product or AI agent with 8,000 app integrations [Web page]. Zapier. https://zapier.com/developer-platform

90. KPMG. (2025, August 19). How AI and automation are changing service delivery. KPMG. https://kpmg.com/us/en/articles/2025/ai-automation-service-delivery.html

91. ElevenLabs. (n.d.). ElevenLabs [AI voice generation platform]. https://elevenlabs.io

92. HeyGen. (n.d.). HeyGen: AI video generator [AI video generation platform]. https://www.heygen.com

93. Synthesia. (n.d.). Synthesia [AI video generation platform]. https://www.synthesia.io

94. AlphaSense. (2025). AlphaSense: Market intelligence and search platform [Web page]. https://www.alpha-sense.com

95. Elicit. (2026). Elicit: AI for scientific research [Web page]. https://elicit.com

96. Clarivate. (2024, December 11). Clarivate launches AI-powered patent search solution in Derwent. Clarivate.

97. Patsnap. (2026, February 23). Need faster patent search? 7 AI-powered tools compared. Patsnap.

98. Cypris. (2026, January 4). The best AI research tools for patent and technical intelligence in 2026. Cypris.

99. McKinsey & Company. (2026, January 12). Gen AI in M&A: From theory to practice to high performance. McKinsey & Company.

100. Black Forest Labs. (2026). Flux [Text-to-image model]. Black Forest Labs. https://bfl.ai

101. Adobe. (n.d.). Adobe Firefly [AI image and content generation model]. https://www.adobe.com/products/firefly.html

102. Getty Images. (n.d.). Getty Images Generative AI [AI image service]. https://www.gettyimages.com/ai

103. Shutterstock. (n.d.). Shutterstock Generative AI [AI image service]. https://www.shutterstock.com/generative-ai

104. Reclaim.ai. (2026). Reclaim: AI calendar for work and life [Web application]. https://reclaim.ai

105. Motion. (2026). Motion: The AI-powered superapp for work [Web application]. https://www.usemotion.com

106. Robert, J. (2026). The impact of AI on work in higher education (Research report). EDUCAUSE. https://library.educause.edu/resources/2026/1/the-impact-of-ai-on-work-in-higher-education

107. Calendly. (2025). Calendly: Online appointment scheduling software [Web application]. https://calendly.com

CHAPTER 13 - The Changing Landscape

108. OpenAI & Penda Health. (2025). AI-based clinical decision support for primary care: A real-world study in Kenya [Preprint]. OpenAI. https://cdn.openai.com

109. Center for Educational Technology. (2023). AI grading systems and teacher workload: Field evidence from K–12 classrooms. (Summary: Third Rock Techkno. (2025). https://www.thirdrocktechkno.com)

110. Li, D. (2025, February 17). Derek Li and Squirrel Ai aim to lead the future of AI-driven education. Forbes China. https://www.forbes.com/sites/forbeschina/

111. Concord. (2025, January 8). AI contract review: The complete buyer's guide. Concord. https://www.concord.app/guide/ai-contract-review-2

112. Xenoss. (2025, July 2). Real-time AI fraud detection in banking: Tools and real-life examples. Xenoss. https://xenoss.io/blog/real-time-ai-fraud-detection-in-banking

113. HubSpot. (2025, November 5). The HubSpot Blog's AI trends for marketers report. https://blog.hubspot.com/marketing/state-of-ai-report

114. Ko, A. (2025, December 13). The authenticity premium: Why consumers are rejecting AI- generated content. Ko Insights. https://www.koinsights.com

115. Koch, J., Meier, L., & Schlegelmilch, B. B. (2024, August 22). The AI-authorship effect. PsyArXiv. https://doi.org/10.31234/osf.io/fq9d7

116. McKinsey & Company. (2024, November 14). Harnessing the power of AI in distribution operations. https://www.mckinsey.com

CHAPTER 14 - The Human Edge

117. Ueda, M., Lucas, R. E., & Fodor, A. (2024). Generative AI enhances individual creativity but reduces the diversity of collective creative output. Science Advances, 10(28), eadn5290. https://doi.org/10.1126/sciadv.adn5290

118. World Economic Forum. (2025, January 7). Rebuilding trust for the age of AI. World Economic Forum. https://www.weforum.org/stories/2025/01/rebuilding-trust-ai-intelligent-age

119. Supply Chain Management Review. (2025, March 3). Reacting to risk: AI's role in supply chain risk management. https://www.scmr.com

120. Gómez, L., & Richter, A. (2024). Analysis of the potential of AI for professional development and talent management. International Journal of Information Management, 78, 102815. https://www.sciencedirect.com/science/article/pii/S2667096824000776

121. Center for Evidence-Based Mentoring. (2025, February 5). New study explores AI and empathy in caring relationships. https://www.evidencebasedmentoring.org

122. Huang, Y., & Patel, R. (2024). Navigating STEM careers with AI mentors. Frontiers in Artificial Intelligence, 7, 1461137.

123. García, M., & Lee, S. (2025). The impact of AI on learning and development: Case studies in corporate training. Journal of Social Sciences, 8(1), 45–62.

124. TechClass. (2026, January 30). Smarter onboarding: How AI is customizing the new hire experience. TechClass. https://www.techclass.com

125. Reis, H. T., Liu, D., & colleagues. (2025). Can generative AI chatbots emulate human connection? Current Opinion in Psychology, 58, 102871.

Additional Resources & Bibliographical References

126. Blue Ridge Global. (2025, July 6). How AI improves forecast accuracy and why it matters. https://blueridgeglobal.com/blog/how-ai-improves-forecast-accuracy

127. 25madison. (2025, October 1). Artificial intelligence in supply chain: From predictive models to AI agents. https://www.25madison.com/content/artificial-intelligence-in-supply-chain-from-predictive-models-to-ai-agents

128. SR Analytics. (2025, October 27). Supply chain predictive analytics: Cut costs 25%. https://sranalytics.io/blog/supply-chain-predictive-analytics

129. Emerj. (2023, December 24). AI for avoiding supply chain disruptions: Two use cases.https://emerj.com/ai-for-avoiding-supply-chain-disruptions-two-use-cases

130. AdSkate. (2025, June 9). AI advertising in 2025: Real campaign results. AdSkate Blog.https://www.adskate.com/blogs/ai-advertising-2025-guide

131. Ampcome. (2025, December 15). AI for supply chain issues: How agentic analytics solves delays, disruptions, and demand shocks. https://www.ampcome.com/post/ai-for-supply-chain-issues-how- agentic-analytics-solves-delays-disruptions-demand-shocks

132. DigiQT. (2025). Supply chain disruption AI agent: Risk management for e-commerce. https://digiqt.com/ai-agent/ecommerce/risk-management/supply-chain-disruption-ai-agent-in-risk-management-for-ecommerce

133. González Vázquez, I. (2025). Comparative analysis of supply chain vulnerabilities and resilience strategies in key sectors (master's thesis). Erasmus University Rotterdam. https://thesis.eur.nl/pub/76401/Gonzalez-Vazquez-Ivonne.pdf

134. Midjourney. (n.d.). Midjourney [AI image generation platform]. https://www.midjourney.com

135. Pfeffer, J. (2010). Power: Why some people have it, and others do not. HarperBusiness.

136. Snaptrude. (2026, March 22). Top 18 AI tools for architects in 2025. Snaptrude. https://www.snaptrude.com/blog/top-18-ai-tools-for-architects-in-2025

137. MIT Technology Review. (2019, August 1). China has started a grand experiment in AI education. MIT Technology Review. https://www.technologyreview.com/2019/08/01/131198/

138. Squirrel AI. (2026). Squirrel AI: Intelligent adaptive learning system [Web page]. https://squirrelai.com

139. HundrED. (2020, June 11). Squirrel AI Learning. HundrED. https://hundred.org/en/innovations/squirrel-ai-learning

140. López, M., & Chen, Y. (2025). The role of AI in shaping the future of education. Cyberpsychology, Behavior, and Social Networking, 28(4), 250–261. https://doi.org/10.1089/cyber.2024.059

141. Park, S., & Ahmed, R. (2025). Artificial intelligence and emotional support. Education and Information Technologies, 30(6), 12345–12368 https://pmc.ncbi.nlm.nih.gov/articles/PMC12785930/

142. Zhang, H., & Müller, K. (2026, March 16). Consumer trust in AI-generated marketing content: A systematic review. American Impact Review. https://americanimpactreview.com/article/e2026024

143. Kang, H. J. (2024). The ethics of using artificial intelligence in medical research. Kosin Medical Journal, 39(4), 321–331. https://doi.org/10.7180/kmj.24.140

144. Rajkomar, A., Hardt, M., & London, A. J. (2024). Recommendations to promote fairness and inclusion in biomedical AI. Journal of Biomedical Informatics, 152, 104338.

145. Bietti, E., & Cohen, I. G. (2024). Co-creating consent for data use. Journal of Law and the Biosciences, 11(2), lsae035.

146. IBM. (2024, January 9). Data suggests growth in enterprise adoption of AI. IBM Newsroom. https://newsroom.ibm.com

147. Marmulstein, J. (2025, June 16). Your "efficient" AI hiring tools might be quietly pushing away your best candidates. LinkedIn.

148. International Monetary Fund. (2026). Bridging skill gaps for the future (Staff Discussion NoteSDN/26/001). IMF. https://www.imf.org

149. Tsan, M., & Rappaport, J. (2026). AI, economic inequality, and the financial hurdles of technological disruption. Technology in Society, 78, 102513.

150. University College London. (2024, July 18). AI boosts individual creativity – at the expense of less varied content [Press release]. https://www.ucl.ac.uk

151. MIT Technology Review. (2024, July 11). AI can make you more creative — but it has limits. MIT Technology Review. https://www.technologyreview.com

152. Tech Brew. (2024, July 15). Study finds AI leads to more sameness in creative writing. Tech Brew. https://www.techbrew.com

153. Li, Y., & Zhao, H. (2025). The homogenizing effect of large language models on creative ideation. Journal of Creative Behavior. Advance online publication.

154. Wang, L., & Chen, Q. (2025). AI-assisted design for cultural and creative products. Scientific Reports, 15, 12345. https://www.nature.com/articles/s41598-024-82281-2

155. Dinçer, Ö., & colleagues. (2025). A multi-stage HR-in-the-loop approach to enhance fairness in AI-based recruitment. IJHRM. https://doi.org/10.1080/09585192.2025.2564235

156. Raghavan, M., Barocas, S., Kleinberg, J., & Levy, K. (2025). Fairness and bias in algorithmic hiring. ACM TMIS, 16(1), 1–36. https://dl.acm.org/doi/10.1145/3696457

157. Singh, P. (2025). AI-driven hiring and its ethical implications. GSC Advanced Research and Reviews, 15(12), 1–12.

158. Nguyen, T., & Silva, M. (2025). Emotional AI and the rise of pseudo-intimacy. AI & Society, 40(3), 455–472.

159. McKinsey & Company. (2025, July 8). The learning organization: How to accelerate AI adoption. McKinsey & Company. https://www.mckinsey.com

160. Harvard Business Review. (2026, March 18). What the best AI users do differently. Harvard Business Review.https://hbr.org/2026/03/what-the-best-ai-users-do-differently-and-how-to-level- up-all-of-your-employees

161. Koh, A. W. L., Lee, S. C., & Lim, S. W. H. (2018). The learning benefits of teaching: A retrieval practice hypothesis. Applied Cognitive Psychology, 32(3), 401–410.

162. Adams, C., Pente, P., Lemermeyer, G., & Rockwell, G. (2023). Artificial intelligence and teachers' professional identity: Mapping the evolution of AI in education toward a co-adaptive and symbiotic future. AI & Society.

163.. Seeley, T. D. (2010). *Honeybee democracy*. Princeton University Press. https://press.princeton.e du/books/paperback/9780691147215/honeybee-democracy

Note: The definitive scientific account of how honeybee swarms reach group consensus through decentralized scout- driven waggle- dance deliberation — the central biological metaphor for collaborative human-AI decision-making throughout the book.

164. Bonabeau, E., Dorigo, M., & Theraulaz, G. (1999). *Swarm intelligence: From natural to artificial systems*. Oxford University Press. https://global.oup.com/academic/product/swarm- intelligence-9780195131598

Note: The foundational text linking insect colony behavior to engineered collective systems provides rigorous theoretical grounding for the book's metaphor of distributed, emergent intelligence that outperforms any individual agent.

165. Hölldobler, B., & Wilson, E. O. (2009). *The superorganism : The beauty, elegance, and strangeness of insect societies.* W. W. Norton. https://wwnorton.com/books/the-superorganism/

166. Preface & Chapter 16: Colony biology metaphor [reference in content]

167. Kegan, R. (1994). *In over our heads: The mental demands of modern life.* Harvard University Press. https://www.hup.harvard.edu/books/9780674445871

Note: Kegan's theory of adult constructive development explains why mid-career professionals often feel cognitively overwhelmed by modern complexity, making it an essential background for the book's diagnosis of the hidden plateau.

168. Arthur, M. B., & Rousseau, D. M. (Eds.). (1996). *The boundaryless career: A new employment principle for a new organizational era.* Oxford University Press. https://academic.oup.com/book/52171

Note: This landmark edited volume coined the term "boundaryless career" and documented the shift away from organizational loyalty toward individual career agency: a structural precondition for the reinvention journeys described in Chapters 2–3.

169. Yang, W., Niven, K., & Johnson, S. (2023). Feeling stuck and feeling bad: Career plateaus, negative emotions, and counterproductive work behaviours. *Human Resource Management Journal, 34*(4), 1031–1049. https://doi.org/10.1111/1748-8583.12539

Note: This three-wave study of 193 UK professionals empirically links both hierarchical and job- content plateauing to negative emotions and counterproductive workplace behavior, providing current evidence (complementary to the Noh 2024 study already cited) for why the hidden plateau matters beyond engagement alone.

170. Clark, D. (2013). *Reinventing you: Define your brand , imagine your future.* Harvard Business Review Press. https://store.hbr.org/product/reinventing-you-define-your-brand-imagine-your-f uture/10835

Note: Clark's practitioner- oriented roadmap for professional reinvention, assessing strengths, reframing narrative, and building a new reputation, provides supplemental how- to depth for the "starting again" chapters.

171. Feiler, B. (2020). *Life is in the transitions: Mastering change at any age.* Penguin Press. http s://www.brucefeiler.com/books-articles/life-is-in-the-transitions/

Note: Feiler's research across 225 life-story interviews identifies five-year "lifequake" transitions as a normative adult experience, substantiating the book's claim that mid-career reinvention is not exceptional but structurally common.

172. Ibarra, H. (2004, September). Career transitions and time horizons. *Harvard Business Review.* https://hbr.org/2004/09/career-transitions-and-time-horizons

Note: Ibarra's research on "possible selves" and the non-linear nature of identity change during career transitions provides academic grounding for the book's framework of trying out new identities before fully committing to them.

173. Epstein, D. (2019). *Range: Why generalists triumph in a specialized world.* Riverhead Books. https://www.penguinrandomhouse.com/books/541546/range-by-david-epstein/

Note: Epstein's evidence that late- specializing generalists often outperform early specialists in complex, unpredictable domains is the primary academic counterargument to hyperspecialization and directly supports the book's advocacy for polymath careers.

174. Ahmed, W. (2019). *The polymath : Unlocking the power of human versatility.* John Wiley & Sons. https://www.wiley.com/en-gb/The+Polymath%3A+Unlocking+the+Power+of+Human+Ve rsatility-p-9781119508489

Note: Ahmed's intellectual history of polymathy from Leonardo da Vinci to modern multi- domain thinkers broadens the context for the book's chapter on why versatility is a strategic advantage in the AI age.

175. Schön, D. A. (1983). *The reflective practitioner: How professionals think in action.* Basic Books. https://www.basicbooks.com/titles/donald-a-schon/the-reflective-practitioner/978078672536

Note: Schön's foundational work on "reflection-in- action" and "reflection-on- action" establishes the theoretical basis for the self- awareness practices the book prescribes for mid-career professionals adapting to new AI tools.

176. Eurich, T. (2017). *Insight: Why we're not as self-aware as we think, and how seeing ourselves clearly helps us succeed at work and in life.* Crown Business. https://www.penguinrandomhou se.com/books/319977/insight-by-tasha-eurich/

Note: Eurich's research shows that only 10–15% of people are genuinely self-aware, despite high self-ratings, directly supporting the book's premise that metacognitive work is a high-leverage, underinvested career skill.

177. Grant, A. (2021). *Think again: The power of knowing what you don't know.* Viking. https://ww w.penguinrandomhouse.com/books/600680/think-again-by-adam-grant/

Note: Grant's research on "rethinking cycles" and the dangers of cognitive entrenchment provides the intellectual humility framework that underpins the awareness and learning pivots described in Chapters 5 and beyond.

178. Deci, E. L., & Ryan, R. M. (2000). The "what" and "why" of goal pursuits: Human needs and the self- determination of behavior. *Psychological Inquiry, 11*(4), 227–268. https://doi.org/10.12 07/S15327965PLI1104_01

Note: The core statement of Self-Determination Theory explaining how autonomy, competence, and relatedness drive sustained motivation provides the theoretical backbone for the book's guidance on setting direction from internal rather than external compulsion.

179. Dweck, C. S. (2006). *Mindset: The new psychology of success.* Random House. https://www.pen guinrandomhouse.com/books/44330/mindset-by-carol-s-dweck-phd/

Note: Dweck's growth-vs.-fixed-mindset framework is indispensable to the book's thesis that mid-career professionals must embrace learning as ongoing rather than finished, especially when confronting AI-driven skill obsolescence.

180. Harford, T. (2011). *Adapt: Why success always starts with failure.* Farrar, Straus and Giroux. h ttps://us.macmillan.com/books/9780374100445/adapt

Note: Harford's argument — drawn from evolutionary biology, economics, and military history — that adaptive experimentation beats top- down planning maps directly onto the book's prescription for "learning by doing" in career pivots.

181. Brown, T. (2009). *Change by design: How design thinking transforms organizations and inspires innovation.* HarperBusiness. https://www.harpercollins.com/products/change-by-desi gn-tim-brown

Note: IDEO CEO Tim Brown's account of design thinking, especially the "bias toward action" and "build to think" principles, provides the creative-process framework for the experimental career moves the book recommends.

182. Peters, T. (1997, August 31). The brand called you. *Fast Company.* https://www.fastcompany.c om/28905/brand-called-you/

Note: Peters's seminal essay, which coined the phrase "brand called you," is the originating text of the professional personal-branding movement and remains essential reading for understanding why Chapter 8's positioning framework matters.

183. Clark, D. (2021). *The long game: How to be a long-term thinker in a short-term world*. Harvard Business Review Press. https://store.hbr.org/product/the-long-game-how-to-be-a-long-term-t hinker-in-a-short-term-world/10512

Note: Clark's research on thought leadership and "strategic patience" shows that sustained visibility, rather than short-term tactics, is the engine of lasting professional reputation, complementing the positioning chapter.

184. Daugherty, P. R., & Wilson, H. J. (2018). *Human + machine: Reimagining work in the age of AI*. Harvard Business Review Press. https://store.hbr.org/product/human-machine-reimaginin g-work-in-the-age-of-ai/10219

Note: Accenture's Daugherty and Wilson document the "missing middle" new roles where humans and AI amplify each other, providing a structured vocabulary for the human-AI collaboration principles central to Chapter 9.

185. Brynjolfsson, E. (2022). The Turing Trap: The promise and peril of human-like artificial intelligence. *Dædalus, 151*(2), 272–287. https://doi.org/10.1162/daed_a_01915

Note: Brynjolfsson's "Turing Trap" thesis that AI focused on mimicking humans reduces wages while AI focused on augmenting humans raises them is the intellectual anchor for the book's preference for augmentation as a career strategy.

186. Bender, E. M., Gebru, T., McMillan-Major, A., & Shmitchell, S. (2021). On the dangers of stochastic parrots: Can language models be too big? In *Proceedings of the 2021 ACM Conference on Fairness, Accountability, and Transparency* (pp. 610–623). Association for Computing Machinery. https://doi.org/10.1145/3442188.3445922

Note: The landmark "Stochastic Parrots" paper identifies the core risks of large language models, including encoded biases and fluent- but-false output, making it an essential background for Chapter 11's failure-mode analysis.

187. Huang, L., Yu, W., Ma, W., Zhong, W., Feng, Z., Wang, H., Chen, Q., Peng, W., Feng, X., Qin, B., & Liu, T. (2023). A survey on hallucination in large language models: Principles, taxonomy, challenges, and open questions. *arXiv preprint arXiv:2311.05232*. https://arxiv.org/abs/2311.05232

Note: This comprehensive survey taxonomizes the causes of LLM hallucination across model architectures and tasks, providing readers with rigorous technical grounding for the practical failure-mode examples discussed in the chapter.

188. Dell'Acqua, F., McFowland, E., Mollick, E. R., Lifshitz-Assaf, H., Kellogg, K., Rajendran, S., Krayer, L., Candelon, F., & Lakhani, K. R. (2023). *Navigating the jagged technological frontier: Field experimental evidence of the effects of AI on knowledge worker productivity and quality* (Harvard Business School Working

Paper No. 24- 013). https://papers.ssrn.com/sol3/papers.c fm?abstract_id=4573321

Note: The BCG field experiment with 758 consultants showing GPT-4 users completed 12% more tasks 25% faster, while tasks "outside the frontier" degraded performance by 19%, is the definitive real-world productivity study for Chapter 12.

189. Peng, S., Kalliamvakou, E., Cihon, P., & Demirer, M. (2023). The impact of AI on developer productivity: Evidence from GitHub Copilot. *arXiv preprint arXiv:2302.06590.* https://arxiv.org/abs/2302.06590

Note: This controlled experiment found developers using GitHub Copilot completed a standardized coding task 55.8% faster than the control group, with especially large gains for less experienced developers, extending the productivity evidence base beyond the writing/consulting studies already cited.

190. Acemoglu, D. (2024). *The simple macroeconomics of AI* (NBER Working Paper No. 32487). National Bureau of Economic Research. https://www.nber.org/papers/w32487

Note: Acemoglu's macroeconomic model concludes AI is likely to raise U.S. total factor productivity by less than 0.66% over ten years, a sober counterpoint to inflated claims, providing Chapter 12 readers with calibrated expectations.

191. Acemoglu, D., & Restrepo, P. (2019). Automation and new tasks: How technology displaces and reinstates labor. *Journal of Economic Perspectives, 33*(2), 3–30. https://doi.org/10.1257/jep.33.2.3

Note: Acemoglu and Restrepo's "task- based" model distinguishes between automation (displacing labor) and new- task creation (reinstating it), providing the economic vocabulary for understanding which jobs AI threatens and which it creates.

192. Autor, D., Mindell, D. A., & Reynolds, E. B. (2022). *The work of the future: Building better jobs in an age of intelligent machines.* MIT Press. https://direct.mit.edu/books/book/5246/The-Work-of-the-FutureBuilding-Better-Jobs-in-an

Note: The MIT Task Force's findings that technology transforms rather than eliminates most occupations and that institutional design matters more than technology itself underpin Chapter 13's nuanced account of AI's labor-market impact.

193. World Economic Forum. (2025). *The future of jobs report 2025.* World Economic Forum. http s://www.weforum.org/stories/2025/01/future-of-jobs-report-2025-jobs-of-the-future-and-the- skills-you-need-to-get-them/

Note: The WEF 2025 report projects 170 million new jobs and 92 million displaced by 2030, with AI literacy among the fastest-growing in-demand skills, directly quantifying the labor-market transformation the book addresses.

194. OECD. (2023). *OECD employment outlook 2023: Artificial intelligence and the labour market.* OECD Publishing. https://www.oecd.org/en/publications/oecd-employment-outlook-2023_08 785bba-en.html

Note: The OECD's cross-national analysis of AI adoption — finding occupations at high automation risk represent 27% of OECD employment provides authoritative data for the book's global labor market perspective.

195. Autor, D. H. (2015). Why are there still so many jobs? The history and future of workplace automation. *Journal of Economic Perspectives, 29*(3), 3–30. https://doi.org/10.1257/jep.29.3.3 Note: Autor's accessible essay demolishes the "lump of labor" fallacy and explains the task- based framework for understanding automation required reading for professionals trying to locate their own skills on the automation frontier.

196. Polanyi, M. (1966). *The tacit dimension.* Doubleday. (Reissued by University of Chicago Press, 2009.) https://press.uchicago.edu/ucp/books/book/chicago/T/bo6035368.html

Note: Polanyi's foundational insight, "we can know more than we can tell," underpins the argument that experienced professionals possess embodied judgment and implicit knowledge that AI systems cannot fully replicate.

197. Vallor, S. (2016). *Technology and the virtues: A philosophical guide to a future worth wanting.* Oxford University Press. https://academic.oup.com/book/25951

Note: Vallor's virtue-ethics framework for living wisely with technology, drawing on Aristotelian, Confucian, and Buddhist traditions, provides the philosophical grounding for the book's case that human judgment and practical wisdom remain irreplaceable in an AI- augmented world.

198. Kotter, J. P. (1996). *Leading change.* Harvard Business School Press. https://www.hbs.edu/fac ulty/Pages/item.aspx?num=137

Note: Kotter's eight- step change model, establishing urgency, building coalitions, and anchoring change in culture, remains the standard organizational change framework and maps directly onto the team-level AI adoption challenges described in Chapter 15.

199. Westerman, G., & Webster, J. (2025, January 22). Generate value from GenAI with "small t" transformations. *MIT Sloan Management Review.*

https://sloanreview.mit.edu/article/generat e-value-from-gen-ai-with-small-t-transformations/

Note: MIT Sloan research showing that companies achieve the most reliable AI value through incremental "small- t" transformations rather than sweeping redesigns, provides a practical adoption framework that supports the book's pragmatic approach to Chapter 15.

200. Crawford, K. (2021). *Atlas of AI: Power, politics, and the planetary costs of artificial intelligence.* Yale University Press. https://yalebooks.yale.edu/ book/9780300252392/atlas-of-ai/

Note: Crawford's field-based investigation of AI's hidden costs in labor, resources, data, and power offers readers a critical ethical counterweight to the book's productivity-oriented chapters, ensuring a balanced view of AI's societal implications.

201. Organisation for Economic Co-operation and Development. (2019). *Recommendation of the Council on Artificial Intelligence* (OECD/ LEGAL/0449). OECD. https://legalinstruments.oecd.org/en/instruments/ OECD-LEGAL-0449

Note: Adopted by 46+ countries and endorsed by the G20, the 2019 OECD AI Principles are the first intergovernmental standard on AI covering inclusive growth, human-centred values, transparency, robustness, and accountability, and remain the baseline vocabulary for nearly every subsequent national AI policy.

202. National Institute of Standards and Technology. (2023). *Artificial intelligence risk management framework (AI RMF 1.0)* (NIST AI 100-1). U.S. Department of Commerce. http s://www.nist.gov/itl/ai-risk-management-framework

Note: NIST's voluntary AI RMF provides a practical four-function framework (Govern, Map, Measure, Manage) for organizations assessing AI risk a concrete institutional supplement to the ethics and governance discussion throughout the book.

203. Benyus, J. M. (1997). *Biomimicry: Innovation inspired by nature.* HarperCollins. https://biomi micry.net/product/janine-book/

Note: Benyus's pioneering book, showing how designers study biological systems (including insect colonies) to solve human challenges, provides the broader biomimicry context within which the book's honeybee metaphor sits.

204. Brooks, A. C. (2022). *From strength to strength : Finding meaning, success, and deep purpose in the second half of life.* Portfolio/Penguin. https://www. penguinrandomhouse.com/books/6 46134/from-strength-to-strength-by-arthur-c-brooks/

Note: Harvard professor Brooks draws on social science and philosophy to argue that transitioning from "fluid" to "crystallized" intelligence in mid-life is a source of strength, a narrative arc that resonates directly with the book's career reinvention thesis.

205. Pink, D. H. (2009). *Drive: The surprising truth about what motivates us.* Riverhead Books. htt ps://www.penguinrandomhouse.com/books/303884/drive-by-daniel-h-pink/

Note: Pink's framework of autonomy, mastery, and purpose explains why mid-career professionals seek meaningful reinvention.

206. Bonelli, R. (2002). *Labor productivity in Brazil during the 1990s* (IPEA Discussion Paper No. 117). Instituto de Pesquisa Econômica Aplicada (IPEA). https://portalantigo.ipea.gov.br/agenc ia/images/stories/PDFs/TDs/ingles/dp_117.pdf

Note: This IPEA analysis documents Brazil's labor-productivity transformation during the 1990s trade liberalization, privatization, and the Plano Real stabilization, providing the macroeconomic backdrop for the author's early career in São Paulo.

207. ECLAC (Economic Commission for Latin America and the Caribbean). (2001). *Brazil in the 1990s: An economy in transition.* United Nations. https://repositorio.cepal.org/handle/11362/ 31749

Note: ECLAC's assessment of Brazil's decade of structural reform rising unemployment, sectoral shifts, and manufacturing globalization, contextualizes the volatile economic environment that shaped the professional experiences recounted in Chapters 1–3.

208. McKinsey & Company. (2024, June). Revolutionizing procurement: Leveraging data and AI for strategic advantage. *McKinsey Insights.* https://www.mckinsey.com/capabilities/operatio ns/our-insights/revolutionizing-procurement-leveraging-data-and-ai-for-strategic-advantage

Note: McKinsey's analysis of how AI is transforming procurement from spend-cube automation to real-time supply-risk modeling, provides a current- state picture of the field the author spent decades in and now sees being reshaped by AI.

209. McKinsey & Company. (2025, October). Transforming procurement functions for an AI- driven world. *McKinsey Insights.* https://www.mckinsey.com/capabilities/operations/our-insig hts/transforming-procurement-functions-for-an-ai-driven-world

Note: "McKinsey projects AI will make procurement 25–40% more efficient, shifting roles from transactional to strategic."

210. Doshi, A. R., & Hauser, O. P. (2024). Generative AI enhances individual creativity but reduces the collective diversity of novel content. *Science Advances*, *10*(28), eadn5290. https://doi.org/10.1126/sciadv.adn5290

Note: "Randomized experiment showing AI boosts individual creativity while making collective outputs more homogeneous."